University Campus Barnsley

IoT

Telephone: 01226 216 885

Catalogue: https://webopac.barnsley.ac.uk

Class No: ....690·068 LAW....................

This book is to be returned on or before the last date stamped below. Thank you!

Labour Conditions for Construction

Labour Conditions for Construction

Building cities, decent work & the role of local authorities

Edited by

Roderick Lawrence

Human Ecology & Environmental Sciences
Faculty of Social and Economic Sciences
University of Geneva

and

Edmundo Werna

International Labour Organization

WILEY-BLACKWELL

A John Wiley & Sons, Ltd., Publication

This edition first published 2009
© 2009 Blackwell Publishing Ltd

Blackwell Publishing was acquired by John Wiley & Sons in February 2007. Blackwell's publishing programme has been merged with Wiley's global Scientific, Technical, and Medical business to form Wiley-Blackwell.

Registered office
John Wiley & Sons Ltd, The Atrium, Southern Gate, Chichester, West Sussex, PO19 8SQ, United Kingdom

Editorial offices
9600 Garsington Road, Oxford, OX4 2DQ, United Kingdom
2121 State Avenue, Ames, Iowa 50014-8300, USA

For details of our global editorial offices, for customer services and for information about how to apply for permission to reuse the copyright material in this book please see our website at www.wiley.com/wiley-blackwell.

The right of the author to be identified as the author of this work has been asserted in accordance with the Copyright, Designs and Patents Act 1988.

Library of Congress Cataloging-in-Publication Data

Labour conditions for construction: building cities, decent work and the role of local authorities / edited by Roderick Lawrence and Edmundo Werna. – first edn.
  p. cm.
  Includes bibliographical references and index.
  ISBN 978-1-4051-8943-9 (hardback: alk. paper)
  1. Construction workers – Employment.   2. Construction industry – Employees – Health and hygiene.   3. Work environment.   4. Labor policy.   5. Quality of work life – Government policy.   6. Local government.   7. Administrative responsibility. I. Lawrence, Roderick J.   II. Werna, Edmundo.
  HD9715.A2L33 2009
  331.2'046900973–dc22   2009001745

A catalogue record for this book is available from the British Library.

Set in 9.5/12.5 pt Palatino by Aptara® Inc., New Delhi, India
Printed in Singapore by Markono Print Media Pte Ltd

1   2009

Contents

Contributors

Professor Yves Flückiger studied sociology and economics at the University of Geneva, where he received his Ph.D. He has been Full Professor at the Department of Economics since 1992. He is also a member of the Board of the Applied Economics Centre (LEA) and of the Employment Centre. He is currently Vice–Rector at the University of Geneva. His scientific interests include employment policies and working conditions in diverse labour markets. He has taught an International Internship Course on Labour and Social Policies to Promote Decent Work, organized by the International Institute for Labour Studies. He also teaches Labour Economics, Industrial Organization and Public Finance. Yves Flückiger has been Research Associate at the Universities of Harvard (USA) and Oxford (UK) and Invited Professor at the Universities of Lausanne, Fribourg (Switzerland) and Deakin (Australia). He has written numerous books and articles published in international reviews (including *The International Trade Journal, Economics Letters, Journal of Econometrics, Journal of Income Distribution, The Oxford Bulletin of Economic and Statistics, Swiss Journal of Economics and Statistics* and *Economie Appliquée*). In recent years, he has received research grants from the Swiss National Foundation for Scientific Research for different projects concerning migrations, gender wage differentials, gender segregation and new forms of employment.

Cedric Lambert trained as a sociologist at the University of Louvain-la-Neuve, Belgium. He has been working as a scientific collaborator at the University of Geneva since 1989. He has conducted small and large-scale surveys on a number of subjects including housing and urban development, using both quantitative and qualitative research methods. He has worked on research grants for the Swiss National Science Foundation for Scientific Research, the Federal Office of the Environment, Forests and Landscapes. He has also completed in-depth analyses of data and statistics for the Geneva Canton and City administrations. He has been an elected municipal officer in the Municipality of Versoix, Geneva, since 2006, and in 2008–2009 he served as the Mayor.

Professor Roderick Lawrence directs the Human Ecology Group at the University of Geneva. He graduated from the Faculty of Architecture and Town Planning at the University of Adelaide (Australia) with First

Class Honours. He has a Masters Degree from the University of Cambridge (England) and a Doctorate of Science from the École Polytechnique Fédérale, Lausanne, Switzerland. In January 1997 he was nominated to the New York Academy of Science. In 1999 he was nominated Professor in the Faculty of Social and Economic Sciences at the University of Geneva. He has been a Consultant to the Committee for Housing, Building and Planning of the Economic Commission for Europe (ECE) in Geneva, and the Urban Affairs Division of the Organization for Economic Cooperation and Development (OECD) in Paris. He is the director of a continuing education course on sustainable development and Agenda 21 at the University of Geneva, which is addressed to administrators in the public and private sectors. His research fields include projects funded by the European Commission, UNESCO and the World Health Organization that consider diverse perspectives and policies for housing, building and urban planning; citizen empowerment and participation; and public and private responsibilities of actors in sustainable development.

Dr Beacon Mbiba is an urban policy expert, with experience both in Africa and the UK. He joined Oxford Brookes University in 2007 as Progamme Leader of the M.Sc. Planning in Developing and Transitional Regions. He was Lecturer in Urban Development Policy in the Department of Geography, London School of Economics and Political Science, LSE (2005–2007). He is a member of the Royal Town Planning Institute and a Fellow of the Higher Education Academy (UK). He has previously taught at the University of Zimbabwe, the Development Planning Unit (DPU) University College London, Birkbeck College, London and the University of Reading. In 2008 Beacon Mbiba was contracted by United Nations (Habitat) to produce a background document 'The State of Cities in Southern Africa' for the UN-Habitat State of African Cities Report 2008/2009, launched at the World Urban Forum, November 2008. In 2004–2005 he worked as Policy Research Analyst and Outreach Executive for The Secretariat to the Commission for Africa (CFA) chaired by former UK Prime Minister Tony Blair. He was co-author of the Zimbabwe National Report and Plan of Action on Human Settlements presented at Habitat II, Istanbul in June 1996.

Dr Michael Ndubiwa is a local government and urban management practitioner with over thirty years of practice in Africa. He was awarded an Honorary Doctorate degree in Social Science by the University of Birmingham in 1995. From 1984 to 1999 he was Town Clerk and Chief Executive Officer of the City of Bulawayo, the second largest city in Zimbabwe and one of the best managed cities in Africa. He has held leadership positions, among others, as Chairman of the Executive Committee of the International Council for Local Environmental Initiatives (ICLEI), Chairman of Zimbabwe Town Clerks' Forum, Chairman of the Municipal Development Program for East and Southern Africa (MDP-ESA) and Board Member of the Reserve Bank of Zimbabwe. Since 2000, Michael has worked as a development consultant and has provided policy advice on programs funded by DFID, GTZ, NORAD covering decentralized local governance, urban water and sanitation, rural and urban linkages.

Mariana Paredes Gil graduated in 2004 with a degree in political sciences from the University of Geneva. She then studied for an M.Sc. in Social and Economic Sciences about 'Globalization and social regulation' from the Universities of Geneva and Lausanne (2006). She served as an intern at the International Trade Center (ITC) in 2001, and at the International Labour Office (ILO) for the InFocus Program on Crisis Response and Reconstruction (IFP/CRISIS) in 2005–2006. She worked as the principal researcher for the GIAN/RUIG international research project 'Promoting Decent Work in Construction: the Key Role of Local Authorities' from 2005 to 2007.

Dr Jill Wells is one of the leading world experts on the construction industry. She is a social scientist/development economist with a degree in economics from the London School of Economics and a Ph.D. in development studies from the University of Wales. She has many years' experience in research and development work focused around the construction sector, with field experience in East Africa (Kenya and Tanzania) and Asia (India, Bangladesh and Vietnam). Her specific areas of expertise are construction industry development and environmental and labour issues related to construction (including work on labour migration, the informalization of employment relationships and occupational health and safety). She has worked as Construction Specialist at the headquarters of the ILO, as a Reader at the Faculty of the Built Environment of South Bank University (London) and as a consultant for international organizations and African governments. She has published widely. Currently, she works at Engineers Against Poverty in London.

Dr Edmundo Werna is an expert in construction and urban development, currently working at the International Labour Organization. His preceding assignment was at the headquarters of the United Nations Volunteers Program, where he designed its urban agenda. Previously, Werna consulted to several development stakeholders, such as the European Commission, World Bank, UN-Habitat, UNCDF/UNDP and WHO. He has also had extensive academic experience, in British, Brazilian and Italian universities and the Woodrow Wilson Center for Scholars (USA). He has published six books, plus many chapters in edited books and several articles in scientific journals. He is also a member of the editorial board of

Habitat International. Werna received prizes such as the Earth Summit 2002 Sustainable Development Award (given by RCA/Stakeholder Forum for Our Common Future), the O. Koenigsberger Award for Best Article of *Habitat International* (1996); honours mention for two articles in the Jorge Hardoy Prize (2000). He co-wrote a chapter of the edited book which won the British Medical Association's Prize for Best Book (1997). One of his WHO projects was chosen as best practice for the Hanover 2000 Exhibition. Two UNV projects also received prizes, and a third one has been selected as Best Practice for the UN-Habitat/Dubai Awards.

Foreword

"The city is a collective space belonging to all who live in it. The inhabitants of cities have the right to conditions which allow their own political, social and ecological development but at the same time accepting a commitment to solidarity. (...) The Municipal authorities encourage, by all available means, respect for the dignity of all and quality of life of the inhabitants".

A total of 354 cities signed this statement as part of the *European Charter for the Safeguarding of Human Rights in the City* adopted in St-Denis, France on 18 May 2000.

The acknowledgement of employment creation as a tool for poverty alleviation is an increasing concern of local authorities. Decent work emerges as an inherent part of social, economic and cultural human rights. Currently UCLG cities are preparing a *World Charter-Agenda of Human Rights in the City* that will contribute to further strengthening the commitment of the local level.

Growing urbanization is resulting not only in a new way to use the planet's resources but also in greater concentration of the labour force and increasing responsibilities for local governments which are the main promoters of infrastructure investments and employment in that sector.

Cost efficient urban development with competitive advantages and high standards are two of the most important challenges to be faced and balanced. Through decentralization in many countries, local authorities increase their role in urban development. They apply construction standards through building codes and promote employment and working conditions through local policy and campaigns.

Currently, globalization challenging the national boundaries of labour policies and political and functional decentralization is changing the roles of local authorities that act simultaneously as employers, regulators and promoters of urban development. This especially affects their relations with migrants, who are vulnerable to changes in the demand of construction-related employment. These factors increase the role of local governments that imply corresponding new responsibilities, including the protection of internationally recognized human and labour rights.

United Cities and Local Governments are the united voice and world advocate of democratic and local self-government. UCLG and its founding organizations have a long history of advocating local governments before the United Nations in all issues of importance for local communities.

This book is a wonderful opportunity to open a new dimension of the debate, acknowledging the key role that local governments need to play in the formulation and enforcement of International Labour Standards in relation to urban development.

UCLG is looking forward to a closer collaboration with the International Labour Organization (ILO), fostering together the exchange of experience in urban employment management and the development of international policy in this field with a strong input from local governments.

Elisabeth Gateau
Secretary General
United Cities and Local Governments

Foreword

Labour is on the agenda of municipal governments throughout the world. I can amply attest to this as a mayor during three terms of office in Diadema[1], an industrial city in the metropolitan region of Sao Paulo, with 400,000 inhabitants and a surface area of 30.7 km^2. I have also had the opportunity to confirm this in my trips to other cities in Brazil and different countries, taking part in local government events, and through liaising with fellow mayors.

Poverty is a major issue in cities and towns throughout the developing world, and this is linked to lack of opportunities for productive and decent employment. It is difficult to think about a way out of poverty that does not entail more jobs and better working conditions.

Urban unemployment and under-employment still abound. However, cities and towns are full of potential and opportunities. Usually, local governments are caught up between the urgency to address poverty, and the challenges to take advantage of the opportunities which could address this problem. In both accounts, local authorities need guidance. The process of decentralization taking place throughout the world, and indeed with an advanced pace in Brazil, has caught many local authorities unprepared, particularly with respect to the issue of labour. Capacity building has been initiated by local authorities to address different issues such as housing, water, sanitation, education and health. However, labour remains a challenge. The need for knowledge and guidance is pressing, and labour in urban areas is still a gap in the policy agenda. Therefore, this book is a timely and important contribution, including analytical material as well as case studies which could inspire and even guide local authorities and other stakeholders.

While providing useful general information on urban labour and the role of local authorities, the book also adds value with its specific focus on the construction sector and related activities. Local authorities have a particular role to play in this sector for several reasons. First, construction is responsible for the very existence and growth of cities and towns. Construction has a great potential for employment creation, but at the same time has many challenges in regard to improvement of working conditions.

In my own experience in Diadema, I would like to mention two examples:

(1)   The development of professional training at the local level. In 1993, at my first term as mayor, I created a professional training school which became a Foundation in 1996. The school offers more than 120 short-term training courses (three to twelve months), some of them in partnership with the productive sector. About 20 of these are related to construction. The novelty is that we had a training course in construction exclusively targeting women.

(2)   A social entrepreneurship program. Since 2003, Diadema has trained small and local entrepreneurs in the area of food and recycling. Tapioca Project targeted small street vendors of tapioca (regional food, like the crêpe in France) and Clean Life Project fostered informal garbage collectors to create a recycling cooperative.

While such initiatives have brought benefits for the workers, further ideas, such as those included in this book, are welcome.

The book also includes case studies in three cities. One of them, Santo André, is close to Diadema. I know well their good efforts. It is good to see their initiatives documented and analyzed in the book. It is also worth noting that the research that led to this book triggered a new concrete project in Santo André, in which the cities of Diadema and Osasco have also collaborated. This is just an example of the application of the knowledge included in the book which, I am sure, will be of value to many local authorities throughout the developing world.

José de Filippi Jr
Engineer, three times mayor of Diadema.

Note

(1)   Diadema was the first city in Brazil (1982) to have a mayor from PT (Workers Party). In 2008, the party had its sixth victory.

Preface

The idea for this book originated from an action-research project carried out by the University of Geneva (UNIGE) and the International Labour Organization (ILO), with funding from the Geneva International Academic Network (GIAN). The editors of the book were also the coordinators of the research project – Roderick Lawrence for UNIGE and Edmundo Werna for the ILO.

Roderick Lawrence was motivated to participate in this international research project because it applied a humanistic perspective that needed to adopt an interdisciplinary research approach. This meant that the project was an extension of his previous research and numerous publications in the field of housing, building and urban planning. He accepted the challenge to work with university colleagues and international civil servants in order to formulate and apply an integrated framework that simultaneously considers the key dimensions of decent work. From the outset he argued that if these key components could not be integrated then the concept of decent work would not be applied; instead it would be another sector based analysis.

Edmundo Werna's motivation came from the fact that the research was an opportunity to combine important aspects of his current work with ILO's increasing interest in the role of local authorities in regard to the 'world of labour'. He also shares Lawrence's enthusiasm regarding a humanistic perspective and interdisciplinarity.

The project started from discussions between the editors/coordinators about the importance of the theme as well as the possible complementarities for working together. UNIGE, being an academic institution, had the qualifications to conduct the investigatory part of the project. The ILO had significant operational experience on the theme, as well as a network which facilitated contacts with professionals to support the research and with the institutions to be investigated. The objective of the GIAN was precisely to promote partnerships between academic and international institutions in Geneva. Therefore, there was a clear common ground shared by the UNIGE, ILO and GIAN, leading to the development of the research, followed by the editing of this book.

The editors would like to thank Randall Harbour of GIAN for all his support as well as technical contributions throughout the whole process. At the ILO, thanks are addressed to Elizabeth Tinoco, Chief of the Sectoral Activities Branch; and Johanna Walgrave, at the time Director of the Social Dialogue, Labour Law, Labour Administration and Sectoral Activities

Department. The Executive-Director, Juan Somavia, is also acknowledged for his promotion of the decent work concept and its application. We hope that the book contributes to this objective. At the same time, Edmundo Werna would like to note that the ILO is not responsible for opinions expressed in this book, and for possible faults and criticisms.

In the completion of the case studies, several people contributed to the research by providing oral information and documents. They are too many to mention here but all have our gratitude. The author of the chapter on Dar es Salaam wishes to thank Eva Lupembe for her support to the research. The author of the chapter on Santo André thanks Alex Abiko for providing preliminary data. The box on Esmeraldas was based on data provided by Jaime Vasconez.

Terms and Abbreviations

AIIPA	Access to Information and Protection of Privacy Act [CAP 10: 27] No. 5/2002. Harare: Government Printer
CABS	Central Africa Building Society
CBA	Collective Bargaining Agreement
CIFOZ	Construction Industry Federation of Zimbabwe
Chimurenga	Revolution or national uprising. Term used to describe uprising of Africans against colonial intrusion in the 1890s (*First Chimurenga*), the armed struggle for national independence in the 1970s (*Second Chimurenga)* and the ongoing land reform, struggle for economic independence and sovereignty (*Third Chimurenga*)
CSO	Central Statistical Office
EIAs	Environmental Impact Assessments
EPZ	Export Processing Zone
GDP	Gross Domestic Product
GoZ	Government of Zimbabwe
Hlalani Khule	Also called *Garikai*, the ongoing housing program following Operation Restore Order
HDI	Human Development Index
ICDS	Inter-census Demographic Survey(s)
ILO	International Labour Organization
IM-LFS	Indicator Monitoring Labour Force Survey(s)
Incidence rate	The number of injuries per 1,000 insured labour force members
Injiva	A status term used to describe Zimbabweans working abroad, especially in South Africa – supposedly 'rich'
Insured labour force	The population at risk of being injured or contracting an occupational disease. Ideally, this should include all working persons. For Zimbabwe, the insured population is taken as that insured under the National Pension Scheme.
LED	Local Economic Development
MDC	Movement for Democratic Change – Opposition Political Party
NAC	National Aids Council

NEC	National Employment Council
NEPAD	New Partnership for Africa
NPS	National Pension Scheme
NSSA	National Social Security Authority
Occupational injury	An injury resulting from an accident during the course of employment
Operation Murambatsvina	*Tsvina* or *Swina* is dirt and Murambatsvina/Murambaswina means he or she who is tidy and does not want dirt. Operation Murambastvina or Restore Order was the government's label for the 2005 program to relocate and reorganize urban informal sector activities.
POSA	Public Order and Security Act [CAP 11: 17] No. 1/2002. Harare: Government Printer
SADC	Southern Africa Development Community
SI	Statutory Instrument
Stand	A term used in Zimbabwe to refer to any undeveloped piece of land or plot reserved for development (industrial, commercial, housing, services, etc.) irrespective of whether or not it is serviced or surveyed.
TNF	Tripartite Negotiating Forum
WCIF	Workers' Compensation Scheme operated by the Ministry of Public Service and Social Welfare
ZANU (PF)	Zimbabwe African National Union (Patriotic Front)
ZAPU	Zimbabwe African People's Union
ZBCA	Zimbabwe Building and Construction Association
ZCATWU	Zimbabwe Construction and Allied Trades Workers Union
ZCTU	Zimbabwe Congress of Trade Unions
ZOHSC	Zimbabwe Occupational Health and Safety Council
ZUCWU	Zimbabwe Urban Councils Workers' Union

Dedication

Roderick Lawrence would like to dedicate his contribution to this book to his sons Xavier, Adrien and Kevin and wishes each of them will have many years of interesting employment ahead.

Edmundo Werna dedicates his work on the book to his daughter, Emiliana, who has been a great source of inspiration even for a book which has not much to do with her universe.

Acknowledgements

This international research project was funded by the Geneva International Academic Network (GIAN-RUIG) and the authors acknowledge this generous support with sincere thanks (http://www.ruig-gian.org). The authors also express their appreciation to all those persons who have provided documentation, made suggestions, or participated in interviews during this project.

Introduction

Roderick Lawrence and Edmundo Werna

What is decent work? Does it differ from a well-paid job, or a safe working environment? Does it deal with conventional characteristics of employment in both the formal and the informal sector? How can local authorities promote decent work? This book answers these and many other questions. The concept of decent work was first presented by the ILO Director-General, Juan Somavia, in his report to the International Labour Conference in June 1999. He stated that the primary goal of the ILO today is to promote opportunities for women and men to obtain decent and productive work in conditions of freedom, security and human rights. Decent work is the converging focus of all its four strategic objectives: the promotion of rights at work, employment, social protection and social dialogue. Thus, the concept of decent work is broadly based on four components: employment generation, social security, rights in the workplace and social dialogue.

This book presents the results of an international applied research project 'Promoting Decent Work in Construction and Related Services: the Key Role of Local Authorities'. The research project analyzed whether the concept of decent work is known and applied by local authorities in the construction sector and related services. This interdisciplinary research project involved a multi-method study of documentary sources and in-depth case studies in Bulawayo (Zimbabwe), Dar es Salaam (Tanzania) and Santo André (Brazil). The book is the product of research carried out by the University of Geneva and the ILO with financial support from the Geneva International Academic Network from November 2005 until November 2007.

Since 1999, the concept of decent work has been promoted and encouraged in different ways and in different regions, countries and economic sectors. Diverse authors have proposed interpretations and indicators in order to better observe the evolution of working conditions and make international comparisons. Therefore, some key questions have emerged from diverse contributions. Is the concept of decent work applicable only to countries with a certain level of economic development and institutional structure, or is it valid across countries with different levels of development, economic structures and socio-economic institutions? What are the goals and main features of the four components of decent work? Is it possible to develop indicators and outline trends towards or away from decent work in different countries?

There have been significant changes in the world since the 1990s. To say the least, there has been a major global shift towards decentralization, giving local authorities much more leverage and responsibility (although not always proportionate to each other) to deal with local matters. Consequently, it is expected that local authorities should play an important role in labour issues, taking into consideration the significant problems related to the generation of new employment and the improvement of existing employment conditions which abound in cities and towns, particularly in developing countries. However, this is still a largely under-researched subject, as are many other topics addressed in this book.

The objective and importance of this book

The contribution of this book is different to other writings on decent work. Although many publications on decent work focus mainly on the roles and responsibilities of actors in the private sector, the contribution of other actors and institutions in the public sector, especially local authorities, should not be ignored. Local authorities do play an important role in economic development, especially through a range of policies and programs in the construction sector and related services (utilities). If empirical research shows that local authorities have made significant contributions to the implementation of decent work, then it is necessary to identify the means and measures that have been applied. On the contrary, if local authorities have not been active, then it is important to understand why they have not been key players in the promotion of decent work. For example, is this related to a lack of understanding of issues, such as significant cuts in public sector budgets combined with increasing trends towards privatization? Or is it related to the fact that the concept of decent work is still unknown or not used by many politicians and employees working in local authorities?

The purpose of this book is to present and analyze the challenges and potential of local authorities to promote decent work by studying the case of the construction industry and related services (utilities). It combines theoretical analysis and case-studies, with recommendations to both practitioners and academics. The research presented in this book is at the interface between knowledge and practice in different disciplines and professions, including:

(1)   Employment policies and labour processes in urban areas (with appropriate inferences from connected fields, including poverty alleviation and social protection against vulnerability).

(2) Municipal management and governance (with appropriate inferences from related fields, such as decentralization and public-private partnerships).

(3) Policies and practices of private companies in the construction industry about their employees' working conditions. These should not be isolated from national employment policies and practices.

(4) Social dialogue, which has facilitated the preceding sets of policies and practices. Social dialogue plays a prominent role in policy making for the creation of decent work. It will also be used in the third phase of this project to communicate the results.

This book is an interdisciplinary contribution addressing urban labour, with a specific focus on the concept of decent work, local authorities and the construction industry. It also makes some inferences to utilities due to their relation to construction. The importance of these topics and their interrelations is elaborated in detail in Chapters 1 and 2. For the purpose of this introduction, these topics are summarized below. These introductory paragraphs also explain the importance of focusing on developing countries.

Labour in urban areas and the concept of decent work

It is already common knowledge that employment generation and improvement in employment conditions are important issues for most elected governments around the world. At the same time, there is ample data on the work-related problems faced by a large part of the population – such as low wages, mistreatment, lack of rights, poor occupational health and safety standards, as well as unemployment or under employment. These problems are particularly acute in developing countries, resulting in an alarming level of poverty. In this context, there are specific reasons for paying attention to urban areas.

Urbanization is a global phenomenon. The twenty-first century has been called the urban century. The urban population of the world has been drastically expanding, both in absolute and in relative terms, especially in the developing world. According to data compiled by UNDP (1999), in 1970 the ratio of city dwellers in developing as opposed to industrialized countries was one to one. Today this ratio is nearly two to one. It will be three to one by the year 2015, and will approach four to one by 2025. Since 1970, 1.23 billion urban residents have been added to the world population, 84% of which have been in less-developed regions. Historically, the rural population throughout the world has been larger than the urban one, but according to UNCHS/Habitat (2006), the year 2007 marked the point in history when both populations became equal. The global trend is that the urban population will continuously outgrow the rural population in the years to come.

Cities and towns are pivotal for development. They are the engines of economic growth. There is a growing amount of evidence about their role in the emerging era of globalization and land markets (Keivani *et al.*, 2001; McGreal *et al.*, 2002; Parsa *et al.*, 2004). They serve as centres for finance and producer services, and are areas of innovation production and the powerhouses of manufacturing and consumer markets. Thus, cities play a crucial role in global, national and regional economies. In sum, there is an economic rationale for investing in urban development.

However, the fact that cities and towns generate significant wealth does not mean that they are devoid of poverty. There are still huge intra-urban socio-economic differentials, as analyzed in detail, for instance, by UNCHS/Habitat (2006) and Werna (2000). While a small proportion of urban citizens in developing countries live in comfort, large numbers still live in poverty. One strong indicator is that there are some 998 million people living in slums in cities throughout the world – one out of every three urban residents (UNCHS/Habitat, 2006). In short, we are witnessing what Nicholas You (2007) describes as the urbanization of poverty. Therefore, there is also a social and ethical rationale for investing in urban development in order to combat poverty. Furthermore, there is an intrinsic connection between poverty and labour. It is difficult to imagine how (urban) poverty can be addressed if the majority of the urban poor work in inadequate conditions, or do not have access to employment at all.

Therefore, there are grounds to take a closer look at labour issues, with a bid to address urban poverty. Decent work is a framework used by the ILO and other organizations to analyze and address issues related to such a theme. It is a comprehensive framework. However, the concept of decent work and its related indicators still has gaps, and these limitations are addressed in this book. It is anticipated that the analysis provided will contribute to strengthening the concept of decent work and its potential contribution to addressing labour related problems.

Local authorities

During the period that globalization has accelerated, decentralization has also become a key concept throughout the world. While there are many variations between countries, both power and management have been decentralized to the local level almost everywhere, in one way or another. The role of local authorities in development has increased, and it has also been reinforced by the creation or strengthening of several associations and networks. Cities and towns are the catalysts of economic growth because they have provided land, infrastructure and services for an increasing share of the global economy.

Therefore, from an economic perspective, it is important to address existing labour-related deficits by promoting a workforce better prepared to face the growing challenges of producing for and competing in a globalizing market. At the same time, there is also a social motivation, that is the scale of urban poverty and its connection to labour-related deficits. For these social and economic reasons, it is reasonable to expect that local government authorities directly involved in the daily management of cities and towns should play a significant role in addressing such labour-related deficits.

However, it is surprising that the role of local authorities in promoting decent work has not received the attention it deserves in recent publications. There have been some general notes, which typically highlight the problems faced by local authorities in fulfilling this role. Therefore, it is necessary to analyse these problems if they are to be overcome. At the same time, it is vital to search for good practices which can be replicated. This book also contributes to that objective.

Construction sector

Construction is vital for human settlements because this industry produces the very fabric of cities and towns. The construction industry is one of the major providers (and in many instances, the primary provider) of work in urban areas, especially in terms of unskilled labour. Further to direct employment on construction sites, the industry provides a range of other jobs, such as the production of building materials and equipment, and post-construction maintenance. It is also worth noting that, in parallel to their jobs in formal building activities, a large number of construction workers supply an important service to other groups of low-income workers through voluntary support and self-help building. For all these reasons, both the quality and quantity of employment in the construction sector have a significant impact on urban development in general and on the alleviation of urban poverty in particular.

Whilst the construction industry is crucial for human settlements, it faces serious challenges related to its workforce in many countries, such as a lack of proper training, high accident rates and large numbers of illegal or unorganized workers. Therefore, there is a good basis for selecting construction to be the subject of an in-depth analysis.

Construction, as opposed to other sectors of the urban economy, also provides a broader scope and potential for local authorities to intervene in labour markets. First, local authorities are playing a strong – and increasing – role in construction, either via the direct execution of public works and/or in some form of partnership with the private sector. In both cases, there is scope for intervention regarding labour, either because construction workers are directly employed by the local authorities, or because such authorities can impose prerequisites on private contracts via procurement regulations. Construction activity is also regulated through planning and building codes, which constitute further instruments for local authorities to address labour issues (although such codes may be limited in regards to labour, they may supplement other measures).

For the above reasons, recommendations made in this book are derived from an analysis of the role of local authorities in promoting decent work in construction. These recommendations are particularly significant for urban development in general and urban poverty alleviation in particular. Some recommendations may also be applied to other sectors or to the urban economy in general.

Related services (utilities)

While the focus of this book is the construction sector, it was also decided to include some information related to utilities, here referred to as 'related (to construction) services', because there is a strong interface between these and the construction sector. First, construction and utilities complement each other as the two basic physical components of human settlements. In addition, utilities and construction often fall under the same administrative rules or unit in a local government. Finally, there is an overlap between the construction workforce and that involved in utilities, particularly in water, sewerage and electricity, especially in the developing world. For the purpose of this book, such services include the provision of water, sewerage and electricity, as well as solid waste management.

Developing countries

Both the theoretical analysis and the case studies provided in this book are focused on developing countries. In these countries and their cities, decent work deficits abound. Therefore these countries must deal with most of the problems and challenges related to the issues addressed in this book. It is also precisely in this context that local authorities face the greatest number of difficulties and challenges. Despite much talk about the benefits of globalization, the construction industry and related services face a number of particular and more pressing problems in developing countries. In sum, it is in the developing world that recommendations can make a difference and where support is needed.

Structure of the book

Chapter 1 presents the concept of decent work. It summarizes numerous definitions and interpretations of it, and presents a study of its four key components. It concludes with a detailed presentation of the methodology used for this research. Each case study has been completed following this methodology and guidelines for field research. The case studies analyze the present situation of each of the selected cities, as well as the national and local contexts, the employment and decent work situation, and, when possible, examples of good practices.

Chapter 2 addresses the general theme of labour. It is organized around the concept of decent work, which is the comprehensive concept used by the ILO and other organizations addressing the 'world of work'. After a presentation of some earlier propositions by ILO specialists to measure decent work, this chapter presents a proposal for the measurement of the four key components using several indicators. This chapter provides background information for the understanding of all chapters of the book.

Chapter 3 considers decent work in the specific context of urban areas, local government initiatives and the construction sector. First, it analyses labour in urban areas. Then, it highlights the evolution of the role of local authorities in the global economy, and specifically in the promotion of decent work, with special attention given to the construction sector and related services. Some examples of good practices concerning the implication of local authorities in employment generation and the promotion of decent work are presented.

Chapters 4, 5 and 6 present the case studies in Bulawayo (Zimbabwe), Dar es Salaam (Tanzania) and Santo André (Brazil). Each case study presents the situation of each of the selected cities, the national and local context, the decent work situation – which has been qualified using decent work indicators – and examples of good practice in the field. The challenges of implementing decent work at the local level are considered. Examples of good practice from other cities are also summarized to illustrate different ways and means of promoting decent work at the local level.

Chapter 7 presents the conclusions, some recommendations and a set of guidelines for international organizations, and national and local authorities concerning the promotion of decent work at national and local levels. These recommendations and guidelines could be reapplied to the construction and other sectors of the economy, especially in developing countries. The conclusion stresses the relevance of decent work as a concept that should be interpreted as a local and global challenge.

1 Conceptual and Methodological Issues

Roderick Lawrence and Mariana Paredes Gil

This chapter traces the origins of the concept of decent work that are associated with some international non-binding agreements. The chapter also presents definitions of decent work used by academics and professionals in the last decade. The content and interrelations of the four key components of decent work are multidimensional and it is noteworthy that conflicts and contradictions between these components have been identified. This chapter also presents criteria for understanding the links between decent work, the construction sector and the initiatives of local authorities. These criteria have been used to define and apply the methodology used in the research which gave origin to this book. The chapter concludes with a description of the research methodology and the availability of sources of information for the three case studies.

1.1   What is decent work?

Decent work was introduced by the ILO Director-General, Juan Somavia, in his first report to the International Labour Conference in June 1999 (ILO, 1999a: 3) using the following words: 'The primary goal of the ILO today is to promote opportunities for women and men to obtain decent and productive work, in conditions of freedom, equity, security and human dignity. Decent work is the converging focus of all its four strategic objectives: the promotion of rights at work, employment, social protection and social dialogue'[1].

Fields (2003: 239) wrote that 'decent work is a new welcome way of achieving the ILO's historic task, for it has shifted the focus to outcomes: what kinds of work people are doing, how remunerative and secure this work is, and what rights workers enjoy in the workplace'. The concept of decent work calls for the creation and promotion of employment, but this should not be the only goal. It also refers to acceptable employment conditions and respect for workers' rights as well as their standard of living.

According to ILO (2002c: 19) decent work is a broad concept of living and working conditions. In general, decent work is a way of guaranteeing human dignity for everyone:

> It is about their job and their future prospects, their conditions of work, the balance between work and family life, the possibility of sending their children to school or withdrawing them from child labour. It is about gender equality, equal recognition and training of women so that they can make decisions and take control of their lives. It is about (their) personal capacity to compete in the market, to keep up to date with new technological skills and stay healthy. It is about developing business skills and receiving a fair share of the wealth they have helped to create and not to be the victim of discrimination. It is about having a voice in the workplace and the community.

A similar kind of humanistic interpretation is proposed by Rodgers (2001: 17): 'Decent work does not refer only to wage employment in large firms. It reflects a broader notion of participation in the economy and the community. It is argued that decent work, rather than just employment or income, should be a basic goal of development, which is equally valid in low income and high income situations.'

Majid (2001: 1) notes that decent work implies the formulation of initiatives allowing the development of the different dimensions of daily life that impact on living and working conditions. It enables a working person to attain a decent life. These dimensions are presented as the strategic components of decent work, which are employment, security, rights in the workplace and social dialogue.

Decent work is an objective common to all societies, which varies according to the customs and conditions of each country. Ghai noted that:

> Working people in all societies desire freedom of association and oppose discrimination, forced labour and child employment in hazardous and harmful situations. They wish to participate through social dialogue in decision-making affecting their work and lives, both at the level of the enterprise and the nation and at the regional and global levels. Likewise, all people in all societies desire work in conditions of dignity and safety and with adequate remuneration. Finally, a modicum of social and economic security in work and life is a universal aspiration.
>
> (Ghai, 2005: 2)

Decent work is a goal. The International Labour Organization (ILO, 1999a: 4–5) emphasizes that 'it reflects in clear language a universal aspiration of people everywhere. It connects with their hopes to obtain productive work in conditions of freedom, equity, security and human dignity. It is both a personal goal for individuals and a development goal for countries.'

The various definitions just briefly presented here confirm that decent work is a much broader concept than the generation of paid work or the simple quantity of work. Decent work means not only the promotion of employment, but good employment conditions with adequate social protection and respect for the rights of workers. Decent work also signifies the promotion of equality among workers and fostering social dialogue.

1.1.1   Origins of the concept

There are some international agreements which preceded the introduction of the concept of decent work in 1999 and which have had an influence on its definition. In particular, the *Universal Declaration of Human Rights* (1948), the *International Covenant on Economic, Social and Cultural Rights* (1966), the *Declaration on the World Summit for Social Development* (1995), and the *Declaration on Fundamental Principles and Rights at Work* (1998) have all provided the foundations for the decent work concept. It is important to highlight some of the principal elements of these non-binding agreements that have inspired the decent work concept.

It is noteworthy that Article 23 of the *Universal Declaration of Human Rights* (1948) states that:

(1) Everyone has the right to work, to free choice of employment, to just and favourable conditions of work and to protection against unemployment.

(2) Everyone, without any discrimination, has the right to equal pay for equal work.

(3) Everyone who works has the right to just and favourable remuneration ensuring for himself and his family, an existence worthy of human dignity, and supplemented, if necessary, by other means of social protection.

(4) Everyone has the right to form and to join trade unions for the protection of his interests.

The idea of employment permitting a convenient standard of living is highlighted in Article 25 of the *Universal Declaration of Human Rights* (1948):

(1) Everyone has the right to a standard of living adequate for the health and well-being of himself and of his family, including food, clothing, housing and medical care and necessary social services, and the right to security in the event of unemployment, sickness, disability, widowhood, old age or other lack of livelihood in circumstances beyond his control.

In 1966, the *International Covenant on Economic, Social and Cultural Rights* pursued the same objectives. In Article 6, it evokes: 'The State Parties to the present Covenant recognize the right to work, which includes the right of everyone to the opportunity to gain his living by work which he freely chooses or accepts, and will take appropriate steps to safeguard this right.'

In particular, Article 7 specifies the objectives concerning employment conditions:

The State Parties to the present Covenant recognize the right of everyone to the enjoyment of just and favourable conditions of work which ensure, in particular:

(1) Remuneration which provides all workers, as a minimum, with:
   (a) Fair wages and equal remuneration for work of equal value without distinction of any kind, in particular women being guaranteed conditions of work not inferior to those enjoyed by men, with equal pay for equal work;
   (b) A decent living for themselves and their families in accordance with the provisions of the present Covenant.
(2) Safe and healthy working conditions.
(3) Equal opportunity for everyone to be promoted in his employment to an appropriate higher level, subject to no considerations other than those of seniority and competence.
(4) Rest, leisure and reasonable limitation of working hours and periodic holidays with pay, as well as remuneration for public holidays.

In 1995, the *Declaration on the World Summit for Social Development*, highlighted the general commitment of 'supporting full employment as a basic policy goal' in its Part C:

We commit ourselves to promoting the goal of full employment as a basic priority of our economic and social policies, and to enabling all men and women to attain secure and sustainable livelihoods through freely chosen productive employment and work.

To this end, at the national level, we will:

(a) Put the creation of employment, the reduction of unemployment and the promotion of appropriately and adequately remunerated employment at the centre of strategies and policies of Governments, with full respect for workers' rights and with the participation of employers, workers and their respective organizations, giving special attention to the problems of structural, long-term unemployment and disabilities, and all other disadvantaged groups and individuals.

These international non-binding agreements form a base for the definition of decent work. The cited norms and rights represent a broad framework for the principles of decent work. Although these non-binding agreements represent some principles for the concept of decent work, they do not constitute all the levels that characterize this multidimensional concept.

1.2   Conceptual issues

Decent work is a concept at the interface between knowledge and practice in different disciplines and professions including:

- Employment policies and labour processes in urban areas, with appropriate linkages to fields including poverty alleviation and social protection against vulnerability (ILO, 2004b).
- Municipal management and governance, with explicit inferences from related fields such as policy sciences concerned with decentralization and public-private partnerships (refer to Chandler & Lawless (1985) for a review).
- Policies and practices of private companies in the construction industry on the working conditions of their employees. These should not be isolated from national employment policies and practices, as noted by Jose (2002).

1.2.1   *Interrelations between the four key components of decent work*

The four key components of the decent work concept are very different and each of them has its own characteristics. However, several contributions show that these characteristics are closely interrelated and that they jointly collaborate in the achievement of common objectives for a whole

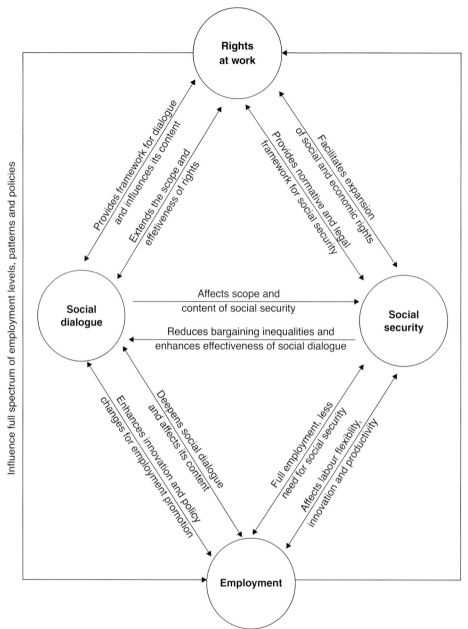

Figure 1.1 Interdependencies between rights at work, employment, social security and social dialogue.
Source: Ghai, D. (2005) *Decent work: Universality and Diversity*. Discussion Paper No. 159, International Institute of Labour Studies, ILO, Geneva, p. 16 (reproduced by kind permission of the ILO).

society (Ghai, 2005; 2006). Following Ghai's analysis, illustrated in Figure 1.1, some features of the interrelations between the four main components of decent work will now be summarized. The first important interdependency issue is how employment levels impact on social security systems. If the level of paid employment in the formal sector is relatively low then this will increase the need for some types of social protection. Thus, both the content and financing of social security systems are influenced by the proportion of workers in each work category. In addition, the distribution of the labour force into different work categories also influences the modes of negotiation between workers unions and employers associations. This affects workers' opportunities and limits their capacity to negotiate other issues such as their rights (Hepple, 2003).

The second point concerning these interrelations is that a developed system of social security can have an effect on labour flexibility and encourage innovation and productivity in employment (Baccaro, 2001). Additionally, a good coverage of the social security system could assist in the development of rights at work, and more broadly in the establishment of social and economic rights. In general, there is a strong relation between social security and rights at and beyond the workplace (Hepple, 2003). In principle, social protection constitutes a strong basis for the extension of social and economic rights.

Third, rights to a minimum wage or a safe work environment have an impact on the form and even the volume of employment (Saith, 2004). In the same way, rights at work (like the right to freedom of association) affect the degree of social protection. With respect to social dialogue, rights at work provide a strong basis for social discussion and also influence the content and purpose of social dialogue.

Finally, we emphasize some interdependencies between social dialogue and the other dimensions of decent work (Kuruvilla, 2003). Social dialogue has a direct impact on the structure and conditions of work. In principle, social dialogue amplifies the sphere as well as the effectiveness and execution of workers' rights. Social dialogue permits negotiations concerning rights at work like social security, minimum wages or adequate conditions of work.

1.2.2 Contradictions and conflicts between components of decent work

Ghai (2005: 15–17) stated that there are two different points of view concerning the conflicts between the key dimensions of decent work. These have been grouped into the interpretations of the neo-classical and institutional schools.

The neo-classical school

The neo-classical school affirms that all state interference in the free performance of market forces leads to inefficiencies in resource allocation and, consequently, to slow growth and wage and employment expansion (except when these are designed to correct market malfunctions). This school maintains that state interventions (for example minimum wages, social security financed through levies on enterprises and collective bargaining) have an undesirable impact on growth, employment and wages owing to discouraging effects on investments, savings and innovations. Additionally, several measures, such as unemployment benefits and welfare payments tend to aggravate unemployment. In developing countries, the neo-classical school argues that the creation of minimum wages and the process of trade unions and collective bargaining have negative economic and social impacts because they concern only a small minority of the labour force. They deform the economies, stress inequalities between working people and restrain investment and job creation. Under these conditions, the introduction of social security systems also has a negative impact

The institutional school

The institutional school argues that despite correcting market failures, state interventions which create rights at work, collective bargaining, tripartite consultations, minimum wages and social security, play a role in political and social stability, reduce economic inequalities, and elevate productivity and innovation. For institutionalists, state interventions tend to alleviate economic fluctuations and enable economic activity and employment to achieve and maintain high levels. Minimum wages and social protection help to develop workers' productivity through better nourishment, superior health and greater security. Trade unions, collective bargaining and tripartite consultations offer an instrument for workers' participation and information allocation, thus increasing mutual trust, sense of responsibility and motivation for better work. In terms of developing countries, the institutional school maintains that due to excessive underdevelopment and mass poverty, the state has to participate in a more substantial way in order to reduce extreme poverty and lower or remove structural barriers to growth.

The viewpoints of these two schools present the major arguments concerning contradictions and conflicts between the four components of decent work. Both interpretations of the neo-classical and institutional schools illustrate the dilemma between efficiency and equity. The dichotomy between the two schools can be observed as a translation of the debate on inequalities in a globalizing economy and the liberalization of markets. The World Development Report 2006 entitled *Equity and Development* stated that while 'some see globalization – greater global integration – as a source of equalization, others see it as a source of widening inequalities, with richer countries and corporations making rules that benefit themselves at the cost of the weak, poor, and voiceless' (World Bank,

2005: 206). It seems difficult to identify the best way to analyze the conflict between the two schools and, of course, between the ways of reacting to globalization and its consequences – specifically what state intervention implies in the employment and other sectors.

It is noteworthy that the two viewpoints presented here highlight the contradictions between the components of decent work. As Ghai (2005: 17) states, a good accord depends on the nature and development of state intervention, as well as labour standards, together with the way in which these components are introduced and expanded. According to the author 'there is, however, widespread agreement that respect of fundamental civil, political, social and economic rights, including core labour standards, is essential for human dignity and indispensable for political stability and sustainable and equitable development'.

Even if we cannot come to any conclusions about the impact of state interventions and the interrelations between the different components of decent work, it seems important to us to insist that interdependencies and/or contradictions should not be presented as a global or normalized issue. Their importance depends largely on national policies and structures as well as on the relative importance attributed to each of the components of decent work.

1.2.3 *Universality of the decent work concept*

Despite the existence of contradictions among the four key components of decent work, the universality of its objectives cannot be challenged. Ghai (2002: 2) stated this as follows:

> All workers, whether in state enterprises, the formal or informal economy or self-employment, desire levels of remuneration in cash or kind that provide at least a minimum standard of living for their families. They also wish to work in safe and healthy conditions and to have a secure livelihood. Like other citizens, workers in all categories also seek the right to form their own organizations to defend and promote their interests and to participate in decisions that affect them as workers.

Nevertheless, as Godfrey (2003: 27) notes, the context in which the main objectives of decent work are considered is different from country to country. The capabilities of this objective will depend on the structure and characteristics of the local and national economy, as well as on the degree of salaried employment, the role of the government and local authorities involved, and the nature of the labour market in each type of sector.

Majid (2001: 2) noted that the accomplishment of a strategic objective for the promotion of decent work is principally measured by the achievement of improving the lives of all. It is evident that the institutional and policy frameworks in which these objectives are developed vary in and between each country and region, according to their own history and customs, and economic and social structure. Therefore, the relative importance given to each decent work objective is best established by each society. Thus, each country, each region and each local authority should define its own decent work agenda and policies according to local circumstances.

1.3   Understanding the interrelationships between employment, construction and local authorities

At the outset, the research envisaged explicit relationships between employment conditions (specifically the promotion of decent work), the construction sector and related services, and the policies and programs of local authorities. These relationships are represented in Figure 1.2. They have not been found in the literature survey of the promotion of decent work, specifically in the construction sector or by the initiative of local authorities. It is anticipated that this framework will provide an innovative theoretical and methodological contribution that will not only enlarge current interpretations of decent work, but also illustrate the pertinence of decentralization and show how the construction sector can contribute to the promotion of decent work and, therefore, to sustainable development at the local level.

Each of the three interrelated subjects shown in Figure 1.2 should be interpreted within the broad economic and political context in which they occur. This context includes national policies and programs in employment and economic development which impact on the construction and other related sectors (education and professional training, housing and community services, social protection and welfare). For example, educational policies that explicitly address continuing education or retaining programs for unemployed workers may or may not be explicitly targeted in the construction sector. Local authorities always implement their

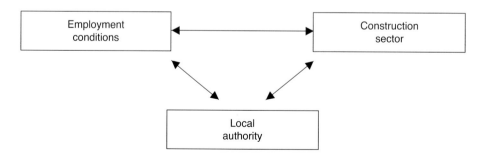

Figure 1.2   Interrelations between employment conditions, the construction sector and local authorities

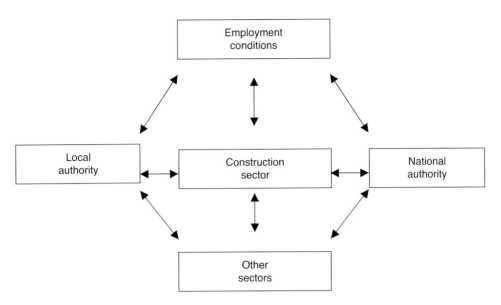

Figure 1.3   Interrelations between employment, construction and other sectors, at both national and local levels

policies and programs within this national context as shown in Figure 1.3. This does not imply that they cannot and do not question national policies and programs. However, if they decide to define different goals and priorities, then this should be documented. This kind of information is crucial and was obtained for each case study. In addition, the policies and programs of local authorities ought to deal with specific local conditions, especially problems such as relatively high unemployment, discrimination or lack of training in the construction sector.

The relationship between local and national policies and programs is dependent on the roles and responsibilities of public authorities at both these administrative levels. Since the 1990s, there has been an increasing willingness to decentralize these roles and responsibilities from the national to the local level. Therefore, in each case study it was important to develop a clear understanding of the scope and limitations of local authorities with respect to employment, and the construction and other sectors. It was also important to compare the construction sector with other sectors in order to identify the specificity of employment conditions in the former.

1.4   Criteria for defining indicators of decent work

A list of indicators adapted to the construction sector has been developed and tested in order to measure the four key components of decent work in either a quantitative or qualitative capacity. These indicators are presented in Chapter 2. They are explicitly related to the four key components of decent work:

- The employment component involves different dimensions such as employment opportunities, paid employment and conditions in the workplace, including safety.
- The social security component concerns social security insurance, old age pensions and others types of social protection for workers.
- The workers' rights component includes forced labour, child labour, inequality in the workplace and freedom of association.
- The social dialogue component involves union density coverage, collective bargaining coverage and other types of social dialogue between employers, government and workers.

1.4.1 Criteria for the construction sector

The construction industry includes both formal and informal production. The relationships between these two sectors can be strong or weak and they may change over time (ILO, 2002a). Data has been collected for the following list of criteria at the national and local levels and, where available, in both the formal and informal sectors:

- Volume of full-time and part-time work
- Volume of shift-work, seasonal work and bonded work
- Level of unemployment
- Salary/wages of workers
- Equity: discrimination against workers (children, women, foreign workers)
- Number of professional accidents
- Level of absenteeism and authorized sick leave
- Profile of workers by age, gender, nationality and level of education/ training

1.4.2 Criteria for local authorities

The performance of local authorities in the promotion of decent work in the construction sector and related services should be evaluated with respect to policies and programs which have been approved by the local administration and/or the national government (Ghai, 2006). Structured interviews have been used to obtain qualitative and quantitative information about:

- Policies and programs for employment in general
- Policies and programs in the construction sector and related services

- Policies and programs on social protection and welfare
- Policies and programs on professional training (apprenticeships and reinsertion)
- Policies and programs on urban community services and infrastructure

The international research project on which this book is based has confirmed the pertinence of analyzing policies, programs and projects in precise sectors at both national and local levels. This is an important finding of this publication because the vast majority of contributions consider decent work only in terms of a broad economic and political framework without reference to sector based initiatives. The exceptions to this customary approach are a limited number of contributions which focus on agriculture. It is important to note that sector-based contributions may be useful in the future to compare the amount of progress made towards decent work in different sectors and also to highlight similarities and differences.

1.5 Research methodology

The research project on which this book is based comprised two main phases which involved a range of methods and collaboration. Phase 1 included a synthesis and critique of recent published contributions on decent work. It also included the definition of criteria and sources of information in order to formulate a set of indicators able to assess decent work in the construction sector in precise localities, based on a synthesis of recent contributions. These are presented in Chapter 2. Phase 2 involved the completion of case studies by field research in three municipalities using the criteria and methodology validated in Phase 1. The case studies evaluated the strengths and weaknesses of policies and programs for the promotion of decent work in each municipality using a set of indicators. This phase was intended to validate the proposed indicators and methodology, which can be reapplied in other localities to measure trends towards or away from decent work. The results of the three case studies are presented in Chapters 4, 5 and 6. A set of guidelines and recommendations which can be reapplied in diverse localities are presented in Chapter 7.

This research protocol applied a combination of quantitative and qualitative methods. An important aim and contribution of the project was to define and validate a set of indicators to evaluate the capacity of local authorities to promote decent work in the construction sector. In order to achieve that goal, it was necessary to define and evaluate criteria about decent work, the construction sector (and related services), and the policies and projects of the local authority in each of the four chosen cities. The findings of this research are presented in Chapters 2 and 3.

1.5.1 *Case studies in local authorities*

The three case studies were respectively in Bulawayo (Zimbabwe), Dar es Salaam (Tanzania) and Santo André (Brazil). Case study methodology has been used in the social sciences and in professional practice throughout the twentieth century (Yin, 1994). A case study is designed to examine the particularity and complexity of a specific case in its naturally occurring setting, using several research methods. A longitudinal perspective may be necessary in order to identify whether or not there have been changes over time.

Our three case studies were proposed in order to apply and validate a conceptual framework and research methodology that could be reapplied in many other situations to study the relationship between employment/decent work, the construction sector, and related policies and programs at both the local and national levels. There is no intention to make statistical comparisons between the case studies. However, even though the context and conditions in these local authorities may be very different, some analytical generalizations have been considered using qualitative and quantitative research methods in order to verify whether or not these relationships are succeeding in each local authority.

The case study for each local authority included three types of empirical research:

- Type 1: the first part of each case study involved the collection and analysis of data and information about employment conditions, specifically in relation to criteria for decent work in the construction sector, at both the national and local levels, and including the formal and informal sectors. Access to data and statistics at both the national and local levels has been acquired for the years 1990 and 2000 in order to have both a comprehensive data set and information about all criteria. (If not available for these years, others could be proposed reasonably close to those years.)
- Type 2: the second part of the case studies was meant to identify links between the data and information obtained in the previous phase and the policies and programs applied in the construction sector at both local and national levels. The key task was to provide a set of quantitative and qualitative indicators for the four dimensions of decent work in both the formal and informal sectors using the criteria and measures established by the project management team.

  In this phase, the empirical research also involved structured interviews with staff from each local authority and representatives of employers' associations and workers unions. A guide to these interviews was developed by the project team in order to follow the same procedure in each of the four selected case studies. These interviews were meant to obtain qualitative information about the goals and outcomes of policies and programs in employment, especially in the construction sector, as well as to discuss examples of best practices and cross-check information provided by the different persons and documents.

- Type 3: the third part of each case study included site visits to projects that were identified as being examples of best practice. Random interviews with companies and workers permitted the verification of the information obtained in the two previous parts of each case study.

1.5.2   Methodology for case studies

The case studies illustrate how criteria can be used systematically to assess the extent to which local authorities are key actors in the promotion of decent work. The case studies have validated the conceptual and methodological frameworks proposed. They have also identified obstacles to the promotion of decent work in the construction sector by local authorities. Some ways and means of overcoming these obstacles have been considered. Specific features of the sources of information used in each case study will be summarized in the following paragraphs.

Bulawayo

A major source of statistical information in Zimbabwe is the Central Statistical Office (CSO). This office was the source for the bulk of statistics on wages, working hours and unemployment. A second key source of statistical information was the National Social Security Authority (NSSA) which collects data on public social security, health and safety, and pensions. Only information considered 'safe' was released for the study, because this information is not for public use. All statutes are sold to the public by the Government printer in Harare and Bulawayo, although some documents can be out of print, in which case one has to depend on libraries such as the one at the University of Zimbabwe. A range of other institutions have sector-based data and written requests were sent to employers and employees' organizations in both Harare and Bulawayo. In addition to the statistics and grey literature, semi-structured, in-depth, face-to-face interviews were conducted with key informants at the CSO, the NSSA, Zimbabwe Congress of Trace Unions (ZCTU), Bulawayo City Council, the National Employment Council (NEC) in Bulawayo and other employers and employees' organizations. The interviewees in Bulawayo included a number of directors in construction firms from the larger organizations, the Construction Industry Federation of Zimbabwe (CIFOZ), as well as from the smaller organizations of indigenous firms, the Zimbabwe Building and Construction Association (ZBCA). Interviews were also held on construction sites with informal sector builders to establish conditions in the sector, with operators at best practice sites (mainly the Kelvin North

Industrial Incubator Shells) and with directors of construction firms involved in the implementation of Bulawayo City's innovative strategies in urban development and decent work.

In the Bulawayo case study, an attempt is made to present statistics and information on decent work indicators, along with a statistical summary of the decent work situation prevailing in the country since the 1990s. Given the need to relate national conditions to local ones, some illustrations are taken from outside Bulawayo, the case study city. Details of the links between local authority policies in urban development and construction and their likely impacts on the decent work agenda in the construction sector were analyzed and are presented in Chapter 4. These also include elements from the limited examples of best practice identified at the local level. Workers' rights, social dialogue and dispute resolution are also discussed. This discussion highlights the legal and political climate within which decent work must be pursued at all levels. Bulawayo City appears to have little input in this sector of decent work outside the area of its legislated jurisdiction.

For the Bulawayo case study, meetings were organized with representatives from the Ministry of Public Service (specifically from the department of Labour and Social Welfare), the National Social Security Authority, the Zimbabwe Congress of Trade Unions, the Zimbabwe Construction and Allied Trades Workers Union, the National Employment Council for the Construction Industry of Zimbabwe, the National Employment Council for Engineering and Iron and Steel Industry, the Zimbabwe Railways Artisans Union, the Zimbabwe Building Contractors Association (ZBCA) and the Zimbabwe National Chamber of Commerce. Other important institutions contacted were the Labour and Economic Development Research Institute of Zimbabwe (LEDRIZ), the Institute of Development Studies and the University of Zimbabwe (namely the Department of Rural and Urban Planning).

Regional and national information was provided by the ILO Subregional Office for Southern Africa in Harare. Additional information was provided by Reliance Construction Pvt Ltd, G.G. Hardware and Construction Pvt Ltd, Belmont Construction Pvt Ltd and Tzircalle Brothers Pvt Ltd. At the local level, the following institutions were contacted: the Mayor and the City of Bulawayo (specifically the Department of Housing and Community Services), as well as the Department of Planning and the Bulawayo Branch of the Zimbabwe Urban Councils Workers Union. Other institutions consulted included the Pumula South Extension, the Kelvin North Industrial Incubator Shells, Cowdray Park and Nkulumane South.

Dar es Salaam

For the Dar es Salaam case study, much of the data has been extracted from the Labour Force Survey (LFS) of 1990/91 (URT, 1993) and the Integrated

Labour Force Survey (ILFS) of 2000/01 (URT, 2001). The results of the two surveys are not always strictly comparable; for example the cut-off point for the inclusion of enterprises in the informal sector rose from five employees in 1990/91 to ten in 2000/01. It should also be noted that data was only available from the labour force surveys for five of the decent work indicators suggested by the study team (unemployment, low wage rate, hours of work, wage inequality between genders and child labour) and only at the national level for all activities combined[2]. There was no regular source of data for the following indicators: working days lost, social security coverage, proportion of the population over 65 not covered by pension schemes, wage inequality according to place of birth, trade union density and collective bargaining. Data from the labour force surveys was therefore supplemented by evidence gleaned from studies undertaken in recent years, as well as from a few ad hoc surveys, notably the Informal Sector Survey of Dar es Salaam (URT, 1998).

Labour Force Surveys were undertaken in 1990/91 and 2000/01 and it was generally possible to make meaningful comparisons between the situations at these two dates and to highlight trends. The Labour Force Surveys (LFS) are national surveys carried out by the central Government. There are no independent local sources of data, but some of the data in the 2000/01 survey was reported separately for Dar es Salaam, although not in 1990/91, so trends in employment in the city are not always detectable. Additional sources of hard data were the Employment and Earnings Survey of 2001, the Tanzania Informal Sector Survey of 1990 and the Dar es Salaam Informal Sector Survey of 1995.

Data is available for most (although not all) of the decent work indicators suggested by the study team. Where statistical data was not available, other sources have been consulted, in particular, two studies commissioned by the International Labour Organization (ILO) investigating the terms and conditions of employment on 11 large construction sites and in the informal construction sector respectively. Other qualitative data collected and analyzed can be considered in three categories:

- Background material covering aspects of Tanzania's political, economic and social development in their forty-five years of independence.
- Materials describing the legal framework for decent work, including a number of Acts of Parliament.
- Documents describing and evaluating some of the donor funded initiatives, past and ongoing, in Dar es Salaam, including a detailed report of working conditions on one of the projects.

Interviews were conducted with key personnel in order to fill gaps in information and ascertain the views of local officials on the responsibility of local authorities for employment generation. Dar es Salaam has four councils: three municipal councils and the Dar es Salaam City Council, which is responsible for coordination. A total of eight interviews were conducted with officials of the Dar es Salaam City Council, but only two with the municipal councils, who proved to be extremely difficult to pin down. Interviews were also conducted with two employer representatives and two workers' organization representatives, as well as with the following: the Chairman of the Hannah Nassif Community Development Association, the project manager of the World Bank Local Government Support Project, the manager of the informal construction workers project; the Chief Technical Advisor of the ILO project 'employment creation in municipal service delivery in East Africa: improving living conditions and providing jobs for the poor'.

All interviews were semi-structured with a short list of key questions, allowing the respondents to talk freely on the subject and to express their views in their own words. The author conducted all interviews, with the assistance of Eva Mbuya from the University of Dar es Salaam, except for those at municipal level, which were conducted by Eva Mbuya alone (and mostly in Swahili).

Additionally, the following institutions were also contacted: the College of Engineering and Technology, University of Dar es Salaam; the UN volunteer, UNV/UNDP/ILO project 'Support to informal construction workers in Dar es Salaam'; the Hanna Nassif Community; the Kinondoni Municipal Council; the Ilala Municipal Council; the Temeke Municipal Council (specifically the Community Infrastructure Upgrading Program) and the National Construction Council, Coordinator of the ILO project 'Baseline study of labour practices on large construction projects in Tanzania'.

At the Dar es Salaam City Council, the following departments or sections were contacted: the Building Section, the Transport Unit and co-ordinator of Dar Rapid Transit (DART), the Community Infrastructure Upgrading Program (CIUP), the Planning Department, the Safer Cities Project and others. Concerning more specifically social dialogue, the following institutions were contacted: the Trade Union Congress of Tanzania (TUCTA); Tanzania Mines and Construction Workers Union; and the Tanzania Mines, Energy, Construction and Allied Workers Union (TAMICO).

Santo André

The objective for the case study in Santo André was to analyze information at both the national and municipal levels for the years 1992 and 2001 in order to study the evolution towards or away from decent work. However, it was found that data on employment conditions for Santo André was not always available. This is a common characteristic of large countries such as Brazil because information is not separated into categories at the municipal level. Sometimes, data was only available at level of the Metropolitan region of São Paulo (MRSP) and the ABC region (which includes Santo André along with six other municipalities). Additionally, statistics specifically related to the construction sector were not easily found, especially at the local level. Sometimes information provided by staff from different departments and secretariats of the Santo André Municipality and by representatives of workers' unions has been used. In this case study, general statistical data has been collected from the Inter-union Department of Statistics and Socio-economic Studies (DIESSE) and the Brazilian Institute of Geography and Statistics (IBGE).

In Brazil, as in the majority of large countries, statistical information has not been separated into categories at the municipal level, but often only at the regional level. In the case of Brazil, this has made it difficult to find published data specifically about Santo André. For that reason, some data has been obtained through interviews with staff of the different secretariats and departments of the Municipality of Santo André. Also, interviews were held with delegates from workers' unions and employers' associations.

The empirical analyses of the interdisciplinary research project applied a methodology that combines quantitative and qualitative information and data. It is divided into three specific parts. The first part involves the collection and analysis of data and information on employment conditions specifically related to criteria about decent work in the construction sector at both the national and the local levels including both the formal and informal sectors for the years 1992 and 2001. The second part is intended to identify links between the data and information obtained in the previous phase and the policies and programs applied in the construction sector at both local and national levels. In this phase, the empirical research also involves structured interviews with staff from different departments and secretariats of the municipal authorities and representatives of employers' associations and workers' unions. These interviews also provide qualitative information about the goals and outcomes of policies and programs in the employment and construction sectors, as well as discussing examples of best practices. The third part of our methodology included site visits to projects that were identified as being examples of best practices.

Whenever possible, the magnitude of the informal sector among the working population has been emphasized. However, it has to be noted that it is extremely difficult to present reliable data about employment in the informal sector, both at national and local levels, given that there are differing interpretations of this sector. It has been particularly difficult to quantify the actual size of the informal labour force, especially in the construction sector.

One of the main problems encountered by the research team was the lack of data and information that could be obtained in a suitable format. For example, not all the relevant information could be obtained for the years 1992 and 2001. In particular, information at the national level frequently dealt with different periods from that available at the local level, making it difficult to compare and gain an understanding of the different research topics.

In the Santo André case study, several people from different departments of the Municipality of Santo André were contacted and met: the Department of Development and Regional Action, the Department of Urban Development and Housing, the Department of Education and Professional Training, the Department of Social Inclusion, the Department of Employment Generation, the Department of Public Services and Works, the Economic Observatory, the Program of Workers' Health and the Public Centre of Employment, Labour and Income (CPETR).

Meetings were organized with the Sacadura Cabral Homeowners' Association, as well as the Brazilian Studio of Art in the Construction Sector. Meetings were also organized with representatives of workers associations, such as the National Confederation of Workers in the Construction Sector (Workers Union in the Construction Sector in Santo André) and the International Federation of Building and Wood Workers (IFBWW), along with representatives of employers associations, such as the São Paulo Union of the Construction Industry (SINDUSCON São Paulo). More statistical data was provided by the Inter-union Department of Statistics and Socio-economic Studies (DIESSE) in São Paulo.

In a more general way, information was also provided by the ILO Regional Office for Latin America in Lima, as well as the ILO Headquarters in Geneva.

Notes

Chapter opening image: Banners showing security rules and prohibiting child labour on a construction site in Bangalore. Photograph courtesy of Marcel Crozet, ILO.

(1)   The term 'productive work' can be understood as referring only to mercantile work and therefore omits all those activities which are not tradable. Basically, there is no specific element which could clarify this topic. According to Juan Somavia, the ILO Director-General (ILO, 1999a: 21), 'without productive employment, the goals of decent living standards, social and economic development and personal fulfilment remain illusory'. Somavia also highlights that there is no 'consensus on the policies most likely to create jobs: for some the issue is one of growth, for others it is labour market flexibility. Some believe that the answer lies in human skills and capabilities, others in policies to share out available work.'

(2)   The exception is data on unemployment, which is shown separately for Dar es Salaam in 2000/01.

2 Measuring Decent Work

Mariana Paredes Gil

This chapter presents a review of numerous proposals to define, classify and measure the four key components of decent work. Each of these components is presented, together with its sub-components and the indicators that have been defined to measure them. At the end of this chapter, Table 2.1 presents a synthesis of all the indicators proposed and tested in the case studies. It is important to note at the outset that, although there is general consensus on the four key components of decent work, the sub-components are not always classified in the same way. For example, working conditions, including health and safety, are classified under social protection in ILO documents, whilst in this book, working conditions have been included as a sub-component of employment: a classification which follows the recommendations made by authors noted in the text.

2.1   Review of proposals to measure decent work

Since 1999, the ILO has published conceptual, empirical and operational studies on decent work. Four main propositions to measure decent work have been suggested by specialists from the different departments and sectors of the ILO in order to study the evolution of decent work in both qualitative and quantitative analyses.

First, Ghai (2003) presented the suitability of different indicators applied to the four major components of decent work: employment, social protection, workers' rights and social dialogue. Each of these four components is represented by a limited number of indicators which can usually be measured in a large number of countries. Accordingly:

> While the first two components of decent work refer to opportunities, remuneration, security and conditions of work, the last two emphasize the social relations of workers: the fundamental rights of workers (freedom of association, non-discrimination at work, and the absence of forced labour and child labour), and social dialogue, in which workers exercise their right to present their views, defend their interests and engage in discussions to negotiate work-related matters with employers and authorities.
>
> (Ghai, 2003: 114)

Ghai states that there are some decent work indicators which are applicable in industrialized countries, economies in transition and developing countries, whereas others are easier to measure either in industrialized countries, or specifically in countries in transition or developing countries only.

Second, Anker *et al.* (2003) introduced a core set of 30 decent work indicators. The authors did not follow the four major components of decent work. Instead, they translated the concept into six dimensions of decent work:

- *Opportunities for work* refers to the need for all persons (men and women) who want to work to be able to find work.
- *Work in conditions of freedom* underscores the fact that work should be freely chosen and not forced on individuals, and that certain forms of work are not acceptable in the twenty-first century.
- *Productive work* is essential for workers to have acceptable livelihoods for themselves and their families, as well as to ensure sustainable development and competitiveness of enterprises and countries.
- *Equity in work* represents workers' needs to have fair and equitable treatment and opportunities for employment. It encompasses absence

of discrimination at work, access to work and the ability to balance work with family life.

- *Security at work* is mindful of the need to help safeguard health, pensions and livelihoods, and to provide adequate financial and other protection in the event of health and other contingencies.
- *Dignity at work* requires that workers should be treated with respect at work, and be able to voice concerns and participation in decision-making about working conditions.

A fundamental element is the worker's freedom to represent his/her interests collectively.

Anker *et al.* note that 'the first two dimensions of decent work – opportunities for work and freedom of choice of employment – are related to the accessibility and adequacy of work. The other four dimensions – productive work, equity, security and dignity – are related to the extent to which accessible and freely admitted work is decent' (Anker *et al.*, 2003: 151–152). In this study, statistical indicators of decent work are considered through the eyes of the general population, in order to recognize the general characteristics and detailed indicators of decent work. Thus, ten groups of indicators are proposed. These are completed by an eleventh group which recapitulates essential aspects of the economic and social context (which describes characteristics of the economy and population that form the context for determination levels and patterns, and the sustainability of decent work). The 11 groups of indicators are: employment opportunities[1], unacceptable work, adequate earnings and productive work, decent hours, stability and security of work, balancing work and family life, fair treatment in employment, safe working environment, social protection, social dialogue and workplace relations, and economic and social context of decent work (Anker *et al.*, 2003: 153–155).

Third, Bescond *et al.* (2003) presented a list based on the 30 indicators proposed by Anker and colleagues. The authors selected seven indicators with data compiled from national labour force surveys conducted in recent years. Their selection measured decent work 'deficits' in terms of low hourly pay, excessive working hours due to economic or involuntary reasons, national unemployment, children not at school (as a proxy for child labour), youth unemployment, the male-female gap in labour force participation, and old age without pension. Bescond *et al.* (2003: 180) highlighted that 'each indicator is introduced, with a brief discussion of its significance, issues related to the quality and international comparability of the relevant data and, where pertinent, alternative approaches.' The authors combined these seven indicators to create a composite index which would measure the decent work deficit at a national level. The chosen indicators are frequently added so that a decent work profile may be constructed by aggregating information for each country. For countries with complete data, the indicators can be included – with or without a weighting according to the population share – in order to acquire an average score for each country as a whole. The resulting single element may be considered as an index of decent work at a specific point in time. Bescond *et al.* (2003: 205) stress the idea that decent work has different meanings for different categories of people:

> For children, decent work means no work at all (or at least no work that conflicts with their schooling). For adults who are currently employed, decent work primarily means adequate pay and no excessive hours of work. For the unemployed, decent work means finding a job (quickly). For the elderly who are no longer economically active, decent work means receipt of an adequate pension from earlier employment. For the young unemployed and for economically active women, an additional consideration is their relative positions with regard to unemployed adults and economically active men, respectively.

Finally, Bonnet *et al.* (2003) have presented a family of decent work indicators applicable at three levels: the macro-level (aggregate), the meso-level (workplace) and the micro-level (individual). This analysis was based on their previous findings (Bonnet *et al.*, 1999). They argued that at the aggregate (macro) level, the purpose of decent work is proposed 'in terms of creating laws, regulations and institutions that enable a growing number of people in all societies to work without oppression, in reasonable security and with steadily improving opportunity for personal development, while having enough income to support themselves and their families' (Bonnet *et al.*, 2003: 213–214). Whilst at the meso-level (workplace), a decent work situation is defined as 'one that provides adequate security for workers while fostering the dynamic efficiency of their enterprises'. Finally, at the micro-level (individual) 'decent work consists of having good opportunity to work with adequate levels of all forms of work-related security'.

The authors suggest seven forms of labour-related security: labour market security, employment security, job security, work security, skill reproduction security, income security and representation security. They propose to create indices based on different combinations of these indicators and thus construct the Socio-economic Security Database. This database consists of five components – three at the macro-level, one at the meso-level and another at the micro-level.

For each security, three types of indicators are required:

(1) *Input indicators* of national and international instruments and rules to protect workers, such as the enactment of basic laws or the ratifications of ILO conventions on work-related hazards, unfair dismissal, the right to organize, etc.

(2) *Process indicators* of mechanisms or resources through which legislated principles and rules are realized, such as public expenditure on a particular form of security, labour inspection services, labour-related tripartite boards, etc.

(3) *Outcome indicators* showing whether or not the inputs and processes are effective in ensuring worker protection. These indicators might include the unemployment rate and the percentage of workers covered by collective agreements or receiving benefits or pensions, etc.

(Bonnet *et al.*, 2003)

For each form of security, three dimensions can be measured (Bonnet *et al.*, 2003: 216):

- The extent to which the government or constitution of the country is committed formally to its promotion.
- The extent to which its institutions give effect to that commitment.
- The extent to which the observed outcomes correspond to reasonable expectations.

The data gathered is used to create the family of decent work indices and the indicators of each form of security are identified. These are then combined to create a composite security index by means of a normalization procedure (based on the one used by the UNDP for the Human Development Index):

$$\text{Normalized value } X = [\text{actual value} - \text{minimum value}]/[\text{maximum value} - \text{minimum value}]$$

Hence:

The actual value is the score achieved by the country on a particular indicator, the minimum value is the lowest value achieved by any country, and the maximum value is the maximum achieved by any country. The average values of all normalized security indices are calculated, and the result is normalized to give values of the decent work index, ranging from 0 (lowest or worst) to 1 (highest or best).

(Bonnet *et al.*, 2003)

The majority of indicators proposed by the different authors can be considered as performance indicators, which measure change over time in response to policy implementation and/or local economic and social situations. These noteworthy proposals collectively represent a broad vision of the measurement of decent work. Our proposition is based on all of them and takes elements from each contribution according to our systemic interpretation of the measurement of decent work. In our opinion, the best way to analyze the decent work situation in a specific country or city (and also to compare different countries or cities) is by using a precise set of indicators for the four key dimensions of decent work. In this way, trends towards or away from decent work can be identified at both national and local levels.

This set of decent work indicators should be composed of indicators which are generally available in a large number of countries, and which come from reliable quantitative databases (such as national surveys and censuses) and qualitative sources (questionnaires and interviews with representatives for different economic actors). In addition, these indicators should be clearly and precisely defined in order to ensure that the same issues are being measured in the different countries or cities. It is also important to define different indicators for the four components of decent work for national or international comparisons. Although creating a decent work index can make comparisons between countries much easier, it may also overlook some locally specific characteristics when combining the indicators and therefore skew the analyses and results.

We prefer to keep some key indicators for each of the four main components of decent work and to make comparisons between these components where possible. In the following sections of this chapter, we will present our proposition for a set of indicators to measure decent work. This list is divided according to the four components of decent work: employment, social security, workers' rights and social dialogue. Where possible, concrete examples of indicators are also presented. The degree to which decent work is applied can be measured using performance indicators for each of the different components. These indicators provide information about the degree of achievement for a particular goal and thus measure the extent to which decent work objectives are met. In addition, indicators may be used to evaluate the performance and progress of these objectives over time, as well as to make comparisons between countries. However, due to conceptual limitations and restricted data availability, it is often impossible to measure an objective exactly. Sometimes an approximate estimation is given. Also, although ideally such indicators should provide a direct measure of a set goal, this is not always possible; in some cases an indirect measurement has to be used. Ghai (2005: 18) stated there is seldom one single indicator that is likely to determine an outcome. Another important element to emphasize is that indicators of the components of decent work can be quantitative and/or qualitative.

The set of decent work indicators proposed in this chapter is based on the four main components of decent work: employment, social security, workers' rights and social dialogue. The goal has been to propose a set of indicators which can be used in all countries. Given that there will be some cases where not all indicators can be measured, we suggest some additional indicators which could be used in order to complete the data. Anker *et al.* (2003: 152) have noted that it is important to keep in mind that any internationally adequate set of ILO decent work indicators will need to be considered as a minimum set. For this reason, we recommend adapting indicators to countries, regions and specific economic sectors in order to be more specific and to identify complementary decent work indicators. In addition, it is important to remember that when international comparisons are made, errors are likely. Bescond *et al.* (2003: 186) noted that these mistakes occur not only because of differences between national data definitions, but also due to the differences in the configuration of accessible data. Egger (2002: 172–173) argued for a step-by-step approach to international comparability using available indicators and data, progressively refining definitions and expanding existing regular surveys or performing new ones.

2.2   Indicators of the employment dimension

The notion of decent work involves the existence of employment opportunities for all those who are available for and looking for work. An important element of decent work is the employment characteristics of a population in a country, region or city. Our proposition follows Ghai's (2003) suggestion that the employment dimension should be apprehended according to three aspects: employment opportunities, remuneration of employment and working conditions.

2.2.1   *Employment opportunities*

According to Anker *et al.* (2002: 9), employment opportunities can be measured in a positive way, in terms of employment and labour force activity corresponding to the population base. Employment opportunities can also be measured in a negative way, in terms of unemployment, underemployment and the absence of employment opportunities. Several authors have noted that there are some indicators which have traditionally been used to measure employment opportunities. These include:

- The labour force participation rate (LFPR)
  Before describing this indicator, it is important to explain how broad our definition of the labour force is. Labour force participants include both employed and unemployed adults. Although not all countries define a legal working age, in this case we include all people aged between 15 and 65 in the working age population, as is the case in a large number of international comparisons.
  To draw on Bescond *et al.* (2003: 200), we will define as 'employed' those people who 'participate in the production of goods and services, if only for an hour during a short, specific reference period

or if they are normally in employment but happen to be away from their work during that period. They may be employers, employees, self-employed workers, domestic helpers, apprentices or members of the armed forces.'

Following the definition of Anker *et al.* (2002: 10), the labour force participation rate measures the level of economic activity of the working-age population in a country. It is a general indicator of the level of labour market activity and gives an outline of the distribution of the economically active population in a country. In other words, the LFPR determines the ratio of the total number of people employed and unemployed with respect to the total resident working age population. Thus LFPR = EPR + UR.

- The employment-population ratio (EPR)

  The employment-population ratio measures the portion of the working age population which is employed. It provides information on the extent to which an economy offers employment in the formal sector. An important advantage of this indicator, suggested by Ghai (2002: 11), is that it provides information on the number and proportion of individuals in the population of working age who are employed in the production of goods and services. In addition, the EPR covers all categories of workers. However, it does not consider the informal sector.

  Nevertheless, Ghai (2002: 11) noted that one of the major sources of change in the EPR across several countries is the participation of women in the labour force. The differences in the extent of the participation definition explain the variations. Thus, in a large number of developing countries, 'women working at home, whether looking after children and the aged or engaged in food preparation, manufacturing, transporting water and wood, or doing repairs are not counted as members of the labour force'. This custom considerably reduces the ratio of employment. Another weakness of the EPR as a measure of work opportunities, noted by Ghai (2002: 11–12), is that it does not give information on the hours worked. Thus, sporadic or self-employed activities are treated in the same way as a work day of eight hours.

  An example of the employment-population ratio is given by the *OECD Factbook* 2005, which presents the EPR for all the OECD countries from 1990 to 2003. In Spain, 51.8% of the working age population was employed in 1990 (EPR = 0.518), whilst in 2003, 60.7% of the working age population was employed (EPR = 0.607), (OECD, 2005: 95).

- The unemployment rate (UR)

  The unemployment rate measures the number of unemployed people as a proportion of the labour force. According to the ILO and Anker *et al.* (2002: 11), a person of working age is classified as unemployed if 'he/she was not employed or has not worked for even one hour in any economic activity (paid employment, self-employment, or unpaid work for a family business or farm), while being available for work, and had taken active steps to seek work during a specified recent period'.

Bescond *et al.* (2003: 190) suggest that national labour force surveys are complete and represent internationally comparable sources of data for measuring unemployment. However, Anker *et al.* (2002: 12) indicate that it is important to use the unemployment rate as an indicator carefully because several aspects of unemployment statistics are not comparable across countries due to methodological differences (e.g. the data source, age group covered, how trainees and other particular categories of workers are counted, and the criteria for deciding what constitutes an active search for paid work). In addition, Godfrey (2003: 8) explains that the unemployment rate can be an awkward measure because the level to which a job-seeker can afford to desist from work (opportunity cost) will vary from one country to another. In industrialized countries this will vary according to the amplitude of the unemployment benefit system, while in developing countries lacking an unemployment benefit system, most of those looking for work must obtain a source of income. Ghai (2002: 13) noted that unemployment rates are often reported to be low in developing countries 'because people cannot afford to stay unemployed'. Thus, 'most potentially unemployed persons either do not actively search for employment, falling into the category of discouraged workers, or they seek a living in the overcrowded informal economy'.

An illustration of the unemployment rate (UR) is given also by the *OECD Factbook* 2005. For the OECD, 'unemployed persons are defined as those who report that they are without work, that they are available for work, and that they have taken active steps to find work in the last four weeks' (OECD, 2005: 108). For example, in Greece 6.9% of the labour force was unemployed in 1990 (UR = 6.9), while 9.3% of the labour force was unemployed in 2003 (UR = 9.3), (OECD, 2005: 109).

2.2.2   *Remuneration of work*

Ghai (2002: 14) notes that an important characteristic of decent work is that workers should benefit from paid employment. This is one element of the quality of work. Anker *et al.* (2002: 23) stress that all individuals who work or seek work do so in order to earn an income and ensure economic well-being for themselves and their families.

Two indicators are proposed to measure remunerated employment:

- Low wage rate
  This indicator is the proportion of the population earning less than half the median wage. Bescond *et al.* (2003: 182–183) note that the

formulation of this indicator – as a percentage of the median wage – makes it independent of national currencies and, therefore, facilitates international comparisons. Another asset of this indicator is its wide applicability, even in countries which have not adopted minimum wage legislation or which have set the legal minimum wage below the current market wage.

- Average earnings in selected occupations
  Anker *et al.* (2002: 26) explain that the choice of specific occupations should be made carefully. Some criteria would be: male-dominated and female-dominated occupations, occupations in tradable and non-tradable sectors, skilled and low skilled occupations.

2.2.3 Working conditions

The range of criteria covered by working conditions may include night work, hours of work, weekly rest, paid leave, and occupational health and safety. In regards to hours of work, we propose two indicators presented by Anker *et al.* (2002: 30): 'excessive hours of work' and 'time-related un-deremployment rate'.

- Excessive hours of work
  The first indicator is measured as the percentage of employed per-sons working more than the threshold number of hours, by status of employment (normally depending on each country). Nevertheless, Anker *et al.* (2002: 31) note that the excessive hours indicator is suscep-tible to a number of factors in addition to real hours worked. These factors include the degree of precision in the measurement of hours worked (which varies from country to country) and the proportion of self-employment in total employment. Sometimes, the case of self-employment is confused with non-economic activities and, therefore, workers in this situation could tend to report higher numbers of work hours. Another important point, highlighted by Anker *et al.* (2002: 31), is that in many cases excessive hours of work occur for economic rea-sons. Most people who combine two or more jobs do so because the income from one job is not enough to maintain their families. It is im-portant to separate excessive hours of work for economic reasons from long or excessive hours of work for other involuntary reasons, such as the nature of work, exceptional circumstances, corporate norms, etc., and voluntary reasons, such as ambition or passion for work. In practice, it may be difficult to make a distinction between excessive hours for 'voluntary' and 'economic' reasons. Therefore, it is impor-tant to exercise due care when applying the excessive hours of work indicator.

- Time-related under-employment rate
  The second indicator used is the time-related under-employment rate, which is measured as the percentage of the employed population working less than a certain number of hours, but who is available and wanting to work additional hours. As Anker *et al.* (2002: 32) explain, time-related under-employment occurs when the 'hours of work of an employed person are insufficient in relation to an alternative employment situation in which the person is willing and available to engage'. This means: willing to work additional hours, available to work additional hours and working less than a threshold number of hours in terms of working time during the reference period[2].

Unfortunately, we have not found any examples of this indicator in practice. However, and in order to illustrate differences in working hours between different countries and sectors, here are some examples of working hour thresholds. According to the ILO, in 2003, 44.8 hours per week were actually worked in the construction sector in Brazil, whereas the average for all economic sectors was 42.2 hours per week (LABORSTA)[3]. In the same year, 43.1 hours were worked in the construction sector in Chile, with a threshold of 43.3 hours calculated for all economic activities. Data was also collected for India, where a threshold of 46.8 hours was worked per week in the construction sector and 46.4 hours were worked in the global economy.

As noted earlier, conditions of work include not only hours of work, but health and safety in the workplace. Anker *et al.* (2002: 49) state that by its very nature, work is exposed to some degree of risk. Risk can appear in several forms, such as repetitive tasks, long hours, exposure to harmful substances, noise, psychological pressure, physical aggression and others. The degree of risk varies according to the occupation, economic activity, type of establishment and characteristics of workers.

In order to measure safety at work, we propose two indicators:

- Number of accidents and deaths at work
  The number of accidents and deaths at work is one widely used indicator measured in relation to the employed population. This indicator can be measured as a proxy per 100,000 employees, for example. Therefore, it is generally known that on construction sites all around the world, the risk of accidents is higher than in other workplaces. According to the ILO, every year at least 55,000 accidents are reported in the construction sector (ILO, 2000c: 4). This implies that an accident occurs in the construction sector somewhere in the world almost every ten minutes. Ghai (2002: 17) notes that in many countries it is hard to obtain information about the consequences of detrimental working environments. However, it is well known that over time 'certain work processes and the use of certain equipment and materials can result in

serious health hazards and diseases'. In addition, disorders related to stress and anxiety are becoming frequent in some working situations.

We will now provide a few examples of workdays lost due to injuries, although there is no data for a proxy of 100,000 employees. As noted by the ILO, 25,098 cases of lost workdays due to injury were registered in the construction sector in Brazil in 2000, while in the same period there were 323,568 cases in the general economy. This means that in Brazil in 2000, 7.75% of injuries with lost workdays happened in the construction sector. In the same year in El Salvador 2038 cases of injury with lost workdays were registered in the construction sector, while 22,845 cases were registered for all economic activity. This means that 8.92% of injuries with lost workdays occurred in the construction sector. The Ecuadorian Institute for Social Security (Instituto Ecuatoriano de Seguridad Social[4]) highlights that for the year 1990, 229 accidents and 23 deaths were reported at the national level. For the year 2000, 138 accidents and 0 deaths were reported at the same level. In this case, data was not available for the construction sector.

- Occupational injury insurance coverage
  In this case we refer to the percentage of employees covered by insurance in relation to injury in the workplace in the construction sector. However, most of the data collected and much of the attention paid to occupational health and safety concerns the formal economy. In contrast, working conditions in developing countries, particularly in the construction sector, tend to be much worse when compared with other types of work (farms, slums, mines, etc.).

Another important element concerning safety at work is the ILO Safety and Health in Construction Convention (167), adopted in 1988. Although this had only been ratified by 23 countries (by February 2009), it is important to note here that this convention exists, because it represents the unique international norm that specifically considers health and safety in the construction sector.

Another important point related to working conditions is the interrelationship between work and family life, emphasized by Anker *et al.* (2002: 38–40). Nowadays, a significant demand for family-friendly work is coming from women, owing to their increasing participation in the paid labour market. Therefore, an indicator can measure the relative importance of paid work and family life:

- Employment rate for women with children under compulsory school-age
  This should be a rate for all women aged 20–49 who have decided to have children and maintain their paid work. Anker *et al.* (2002: 39) state that 'official labour force statistics classify women with young children as employed, whether they are at work or on maternity leave with the

expectation of returning to the same employer'. The authors suggest we be aware of those cases where women are forced to go back to work for economic reasons. This kind of situation should not be considered as decent work, since women do not have the option to go back to work.

- Paid leave
  Another indicator used to analyze conditions of work should be the number of days of paid leave. Bonnet *et al.* (2003: 221) propose that this indicator can be measured as the average of annual paid leave (holidays), adjusted for the portion of workers in formal paid employment.

- Comfort and hygiene conditions
  This indicator is difficult to measure, because there is not much official information in a number of countries. Comfort and hygiene conditions should be measured as a qualitative indicator referring to the presence of a comfort and hygiene infrastructure in workplaces, including a first aid room, staff room, canteen, showers and toilets.

2.3   Indicators of the social security dimension

According to the ILO, social security 'encompasses the broad areas of income security, health protection and the assurance of safe and healthy working conditions, all of which are closely interrelated' (ILO, 2003a: 52). Ghai (2002: 17) noted that social security systems were proposed 'to meet people's urgent subsistence needs and to improve protection against contingencies'. The crucial problem is that these systems were generally meant only for salaried employees in the formal sector.

Anker *et al.* (2002: 52) stressed that the measurement of social security should try to quantify the extent of coverage offered for its three principal aspects: population coverage (in terms of access, entitlements and contributions), benefits levels and expenditures. Therefore, we propose the two following indicators for measuring the social security dimension of decent work:

- Public social security expenditure
  This indicator measures the public resources devoted to social security as a percentage of the GDP (statistics for total social security, health services and old-age pensions). However, as Ghai (2002: 18) notes, this indicator does not concern 'the efficiency with which these resources are used'. Ghai (2005: 17–19) insists on differentiating between industrial and transition countries on the one hand, and developing countries on the other, in order to measure the indicators of social security effectively. In industrial and transition countries, the ratio of public social security expenditure to GDP provides a good indicator of the coverage

and level of benefits accorded to workers. Nevertheless, as previously noted, in developing countries, where a large part of the working population is employed in the informal sector, these social security systems are insufficient to meet the essential needs of the population.

- Old age pension

    Bescond *et al.* (2003) argue that this indicator measures the number of people who receive a pension as a result of their past economic activity. However, and in order to have a better indicator of this subject, the authors suggest separating those 'persons in receipt of a pension or income from a property or investment', from those 'persons who receive a pension based on their past economic activity' (Bescond *et al.* 2003: 203).

Anker *et al.* (2002: 55) suggest another important, but not readily available indicator: health insurance coverage. Saith (2004: 28) emphasizes a significant lack of social security indicators; a poor country may have a very low rate of formal pension coverage, but in reality have good quality elderly care provided by family members. The system may even be better than in developed countries, where family ties are weaker and there is less financial dependence on family members. In this type of system, a greater rate of pension coverage may imply better old age care. For this reason, it is important to establish the need for social protection and the extent of insecurity, and then measure the coverage of requirements specific to each society. An essential element of the social security dimension is the coverage of the national social security system as well as all its characteristics (such as the population and spheres covered, and the system reform, if any). Thus, even if this qualitative element does not facilitate comparisons among countries (for obvious reasons) it seems important to improve understanding of the social security systems in each country, region or city being studied.

In principle, although the establishment of social protection systems distinguishes development objectives in many countries, in reality the majority of social security structures do not correspond to the needs of workers in some economic sectors, especially in the construction sector. Generally, the health of workers in industrial sectors is determined to a broad extent by the quality of their shelter, water supply, hygiene conditions and health services, as well as their earnings and poverty levels. In other words, 'the deficits in decent work, such as low and fluctuating income as a result of uncertain and insecure employment lead to extreme vulnerability among workers when income is lost' (ILO, 2003a: 52). Days lost from work as a result of work-related injury, disease, invalidity, death or natural disasters generate deficiencies among workers. These workers have to confront social and health risks occurring in the hazardous and unstable situations characteristic of many industrial sectors.

2.4   Indicators of the workers' rights dimension

Decent work must be work that is acceptable to society. It is therefore necessary to know the incidence of unacceptable work, both to ensure that such work is excluded from indicators of employment opportunities and also to measure progress towards its elimination. The main concern here is with forced labour, child labour under abusive conditions, discrimination at work and freedom of association.

2.4.1   Forced labour

In 1930, the ILO Convention No. 29 concerning 'Forced or Compulsory Labour' defined forced labour as: 'all work or service which is exacted from any person under the menace of any penalty and for which the said person has not offered himself voluntarily'. Forced labour is very difficult to measure precisely. As Anker *et al.* (2002) explain, it is hidden since it is illegal and immoral. Thus, it is not just difficult to measure but 'it would be systematically under-reported using typical household or enterprise surveys' (Anker *et al.* 2002: 20–21). It is therefore very hard to find a good indicator of forced labour as a constituent of decent work. An alternative can be found using indirect qualitative measures of forced labour using ILO studies and reports on human rights abuse.

2.4.2   Child labour

According to the UNICEF Report *The State of the World's Children 2006* and following the latest estimations from the ILO, '246 million children between 5 and 17 are engaged in child labour. Of these, nearly 70% or 171 million children are working in hazardous situations or conditions, such as mines, with chemicals and pesticides in agriculture, or with dangerous machinery' (UNICEF, 2005: 46).

There is no official minimum age standard for working. Each country can freely choose an age, according to its own law. However, the ILO Convention No. 138, Article 1, published in 1973, specifies that 'Each member for which this Convention is in force undertakes to pursue a national policy designed to ensure the effective abolition of child labour and to raise progressively the minimum age for admission to employment or work to a level consistent with the fullest physical and mental development of young persons.'

Article 2 of the Convention states that:

(1) Each member . . . shall specify, in a declaration appended to its ratification, a minimum age for admission to employment or work within its territory . . .
(3) The minimum age . . . shall not be less than the age of completion of compulsory schooling and, in any case, not be less than 15 years.

Another important convention concerning child labour is the *Convention on the Rights of the Child*, adopted by the UN General Assembly in 1989, which stipulates in Article 32:

(1) State Parties recognize the right of the child to be protected from economic exploitation and from performing any work that is likely to be hazardous, to interfere with the child's education or to be harmful to the child's health or physical, mental, spiritual, moral or social development.
(2) State Parties shall take legislative, administrative, social and educational measures to ensure the implementation of the present article. To this end ... State Parties shall in particular:
  (a) Provide for a minimum age or minimum ages for admission to employment;
  (b) Provide for appropriate regulation of the hours and conditions of employment;
  (c) Provide for appropriate penalties or other sanctions to ensure the effective enforcement of the present article.

Ghai (2002: 20) stated that child labour is widespread in developing countries in different forms, such as work on family farms or enterprises. It is important to take care when distinguishing the circumstances where child labour is harmful to the health and well-being of children and to their future prospects. These situations include children working in mines, factories and on construction sites, as well as work in dangerous and unhealthy environments. Worst of all is the exploitation of children in sexual commerce, forced or bonded labour, armed conflict and human trafficking.

Following the ILO Convention No. 182, Article 2, published in 1999, the worst forms of child labour include:

(1) all forms of slavery or practices similar to slavery, such as the sale and trafficking of children, debt bondage and serfdom and forced or compulsory labour, including forced or compulsory recruitment of children for use in armed conflict;
(2) the use, procuring or offering of a child for illicit activities, in particular for the production and trafficking of drugs as defined in the relevant international treaties;
(3) work which, by its nature or the circumstances in which it is carried out, is likely to harm the health, safety or morals of children.

Child labour is particularly significant in some specific sectors, including the construction sector. UNICEF noted that 'one boy in every four and more than one in every three girls working in construction suffers work-related injuries and illness' (UNICEF, 2005: 47).

Anker *et al.* (2002: 18–19) suggest two indicators to measure child labour:

- Children not attending school
  The percentage of children not attending school (percentage by age) is a good proxy measure for unacceptable child labour, as well as being a useful indicator and goal in its own right for child welfare. According to UNICEF in its *State of the World's Children 2006*, the ratio (net) of primary school attendance in Namibia in 2004 described only 78% of children of compulsory primary school age (UNICEF, 2005: 114). This means that 22% of Namibian children of compulsory primary school age do not go to school, and that they are probably working.
- Children in wage employment or self-employment activity rate
  Children working as employees or in self-employment are a second proxy indicator for unacceptable child labour. This often occurs under exploitative conditions and is detrimental to children's health, safety and morals. This interferes with school attendance and educational performance and this in turn decreases lifetime employment options. These forms of child labour can damage children physically and/or mentally. They also prevent children from going to school, therein harming the productive capacity of the future workforce. According to UNICEF, in 2004 child labour affected 23% of children in Vietnam and 14% in India (UNICEF, 2005: 130).
  It should be noted that for UNICEF, child labour represents the percentage of children aged 5 to 14 years of age involved in child labour activities at the moment of the survey. A child is considered to be involved in child labour activities under the following classification:

(1) Children 5 to 11 years of age that during the week preceding the survey did at least one hour of economic activity or at least 28 hours of domestic work.
(2) Children 12 to 14 years of age that during the week preceding the survey did at least 14 hours of economic activity or at least 42 hours of domestic work.

Figure 2.1 shows the worst forms and relative importance of child labour in the world. Hence, the majority (59%) of the worst forms of child labour represent children in forced and bonded work, followed by children in prostitution and pornography (19%). Together they constitute almost 80% of the worst forms of child labour.

According to the ILO (2002d: 35–36), the following definitions are counted as unconditional worst forms of child labour:

(1) Trafficked children: based on the UN Protocol to Prevent, Suppress and Punish Trafficking in Persons, especially women and children, child trafficking is defined as 'the recruitment, transportation,

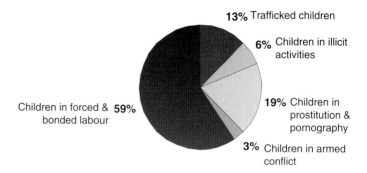

Figure 2.1   Children in unconditional worst forms of child labour
Source: Own composition, based on ILO (2002d) *Every Child Counts: New Global Estimates on Child Labour,* International Programme on the Elimination of Child Labour (IPEC) and Statistical Information and Monitoring Programme on Child Labour (SIMPOC), International Labour Office, Geneva. p. 25.

transfer, harbouring or receipt of a girl or boy of less than 18 year of age for the purpose of exploitation'. In this case, this is limited to children trafficked for sexual and economic exploitation, including '(a) child prostitution; (b) other forms of sexual exploitation such as the use of children for pornography; and (c) forced labour or services, slavery or practices similar to slavery and servitude'.

(2) Children in illicit activities: in this case, reference is made to the ILO Convention No. 182 which refers to children in illicit activities as 'children involved in the production and trafficking of drugs'. Following this emphasis, this case therefore includes 'children in drug manufacture, including work in poppy plantations and trafficking of illegal substances'.

(3) Children in prostitution and pornography: according to the Optional Protocol for the Convention on the Rights of the Child on the sale of children, child prostitution and child pornography, adopted by the UN General Assembly on 25 May 2000. Child prostitution includes 'the use of a child in sexual activities for remuneration for any form of consideration'; and child pornography refers to 'any representation, by whatever means, of a child engaged in real or simulated explicit sexual activities or any representation of the sexual parts of a child for primarily sexual purposes'.

(4) Children in armed conflict: this refers to 'any person under 18 years of age who is part of any kind of regular or irregular armed force or armed group in any capacity, including cooks, porters, messengers, and those accompanying such groups, other than purely as family members. It includes children recruited for sexual purposes and forced marriage.'

(5) Children in forced and bonded labour: 'forced child labour can be distinguished from the other forms of child of labour through the presence of one or more of the following elements: (a) a restriction of the freedom to move; (b) a degree of control over the child going beyond the normal exertion of lawful authority; (c) physical or mental violence; and (d) absence of informed consent'.

2.4.3   Inequality at work

Inequality at work is a characteristic that should be eradicated if acceptable workers' rights are to be achieved. The *World Development Report 2006* proposed the measurement of inequalities in different spheres, such as inequalities in health, inequalities in education, and economic inequalities. It stated that when we talk about inequalities we are referring to: 'systematic differences in opportunities for individuals and groups who differ only in skin colour, caste, gender or place of residence; predetermined characteristics that can be argued as morally irrelevant' (World Bank, 2005a: 28). Inequality is difficult to measure, because it is a qualitative characteristic of human groups and therefore variable between cultures and societies. For these reasons, we propose to analyze discrimination instead of inequality.

In its policies to reduce inequalities at work, the ILO emphasizes discrimination at work which involves the denial of equality of treatment and opportunity to individuals in their own right or as members of a social group. According to the ILO Convention No. 111, Article 1 (1958), discrimination comprised: 'any distinction, exclusion or preference made on the basis of race, colour, sex, religion, political opinion, national extraction or social origin which has the effect of nullifying or impairing equality of opportunity or treatment in employment or occupation'.

In 1995, the Beijing Declaration from the Fourth World Conference on Women stipulated Government participants' commitment to promote:

> The equal rights and inherent human dignity of women and men along with other purposes and principles enshrined in the *Charter of the United Nations*, the *Universal Declaration of Human Rights* and other international human rights' instruments, in particular the *Convention on the Elimination of All Forms of Discrimination against Women* and the *Convention on the Rights of the Child*, as well as the *Declaration on the Elimination of Violence against Women* and the *Declaration on the Right to Development*.

In addition, the elimination of discrimination at work also means fair treatment. This implies working without harassment or exposure to violence, some degree of autonomy, fair handling of grievances and the resolution of conflicts. The most commonly used discrimination indicators concern gender, but in principle other similar kinds of indicators may also be used. As the ILO Global Report *Time for Equality at Work* stated 'whatever the form of discrimination – be it based on race, sex, age, disease or disability – its elimination tends to require a similar set of policy devices. These range from consistent and effective regulatory and institutional frameworks to suitable training and employment policies' (ILO, 2003c: xii).

According to the ILO Global Report, 'discrimination in employment and occupation takes many forms, and occurs in all kinds of work settings' (ILO; 2003c: 1). In a general way, 'to discriminate in employment and occupation is to treat people differently and less favourably because of certain characteristics, such as their sex, the colour of their skin or their religion, political beliefs or social origins, irrespective of their merit or the requirements of the job' (ILO, 2003c: 15). Moreover, all types of discrimination have one common feature: it 'entails treating people differently because of certain characteristics, such as race, colour or sex, which results in the impairment of equality of opportunity and treatment' (ILO, 2003c: 1).

The difference between direct and indirect types of discrimination at work needs to be considered. Following the ILO Global Report concerning discrimination at work, direct discrimination occurs 'when regulations, laws and policies explicitly exclude or disadvantage workers on the basis of characteristics such as political opinion, marital status or sex' (ILO, 2003c: 19). Contrarily, indirect discrimination 'may occur when apparently neutral rules and practices have negative effects on a disproportionate number of members of a particular group irrespective of whether or not they meet the requirements of the job'. Thus, 'indirect discrimination may also occur when differential treatment is accorded to particular categories of workers' (ILO, 2003c: 20–21).

The *Convention on the Elimination of All Forms of Discrimination Against Women*, endorsed in 1979 by the UN General Assembly, stated in its Article 11:

> State Parties shall take all appropriate measures to eliminate discrimination against women in the field of employment in order to ensure, on a basis of equality between men and women, the same rights, in particular:
>
> (1) The right to work as an inalienable right of all human beings
> (2) The right to the same employment opportunities, including the application of the same criteria for selection in matters of employment
> (3) The right to free choice of profession and employment, the right to promotion, job security and all benefits and conditions of service and the right to receive vocational training and retraining . . .

(4) The right to equal remuneration, including benefits, and to equal treatment in respect to work of equal value, as well as equality of treatment in the evaluation of quality of work

(5) The right to social security, particularly in cases of retirement, unemployment, sickness, invalidity and old age and other incapacity to work, as well as the right to paid leave

(6) The right to protection of health and to safety in working conditions, including the safeguarding of the function of reproduction

Our proposition of indicators for discrimination is related to gender. However, these indicators can also be used to measure discrimination at another level or between different groups of workers. Four indicators may be used to measure gender discrimination:

- The male-female gap in labour force participation
  Bescond *et al.* (2003: 200–201) noted that the male-female gap in the labour force participation rate is the desegregation by sex and by age of the labour force participation rate (LFPR). An indicator of the male-female gap in labour force participation measures the extent to which women enter the labour market relative to men across different countries. The authors remind us that this gap may be narrowing in some countries due to a decline in male participation, not because of an increase in female participation in the labour force.

- Unemployment rate for men and women
  Generally speaking, unemployment rates have always been higher for women than for men. As the ILO Report *Time for Equality at Work* (2003c: 42) noted, 'women, and for that matter other discriminated-against groups, may adjust to deteriorating labour market conditions by accepting shorter working hours rather than no work at all, and therefore become underemployed and, in the face of discrimination, become discouraged and abandon active job seeking altogether.' The ILO Report also specified that there are differences between men and women in terms of employment status. Thus, men often have regular and more highly remunerated positions, while women 'are often in peripheral, insecure, less-valued positions'. In addition, women represent the majority of home workers, casual workers and temporary workers (ILO, 2003c: 42).
  To illustrate the gap between the unemployment rate for women and men, we present statistics from the *Panorama Laboral 2005* published by the ILO Regional Office for Latin America and the Caribbean in 2005. According to the ILO, during the first semester of 2005 the urban unemployment rate in Colombia represented 15% of the total labour force population, but this affected women and men in different ways. For women, the urban unemployment rate for the first semester 2005

described 17.6% of the female labour force, whereas it concerned only 12.6% of the male labour force during the same period (ILO, 2005a: 92).

- Differences in earnings

  Some authors highlight the difficulties in defining a precise indicator for measuring differences in salaries between men and women owing to their concentration in different occupations and economic sectors, as well as differences in the number of hours worked.

  For the ILO, discrimination in paid wages is an important illustration of what discrimination at work means. The ILO (ILO, 2003c: 47) clarified that 'discrimination in remuneration occurs when the main basis for the determination of wages is not the content of the work to be performed, but rather the sex, colour or personal attributes of the person performing the work.' In order to really examine discrimination in remuneration, the ILO has presented two proposals. The first 'distinguishes between inequalities due to individual characteristics such as levels of education, skills, or seniority, and inequalities due to discrimination'. The second proposition 'focuses on inequalities amongst groups and finds that women's earnings in certain occupations, sectors, skills or levels of pension are typically lower than men's, irrespective of individual abilities' (ILO, 2003c: 47).

- Distribution of skilled jobs

  The distribution of men and women across levels of responsibility is an important measure of equal treatment in employment. One indicator of this is the extent to which women are employed in positions of authority and decision-making, such as managerial and administrative positions.

2.4.4 Freedom of association

Freedom of association is necessary for workers and employers to protect their interests, to coordinate common activities and to participate in negotiations and dialogue concerning their interests. Following Ghai's contribution (2002: 24–26), two quantifiable indicators are proposed:

- The number of countries having ratified Conventions No. 87 and 98

  Convention No. 87 refers to 'freedom of association and protection of the right to organize' and Convention No. 98 refers to the 'right to organize and to bargain collectively'. However, ratification of a Convention does not necessarily guarantee that the country that signed it respects the necessary conditions for freedom of association.

  In 2007, according to the ILO, Convention No. 87 had been ratified by 145 countries out of the 178 members of the Organization, and 154 countries had ratified Convention No. 98. Violations of freedom of association among workers in the construction sector are common. These abuses vary from regulatory restrictions and anti-union practices to physical violence and even repression.

- Index of civil rights

  One of the most frequently used indices of civil rights is the civil liberties index, prepared by Freedom House. This index is based on comprehensive subjective evaluations by human rights experts who collect evidence globally. The civil liberties index is based on a checklist including: freedom of expression and belief, association and organizational rights, rule of law, personal autonomy and individual rights.

  According to Freedom House, each country and territory is assigned a numerical rating, on a scale of 1 to 7. A rating of 1 indicates the highest degree of civil freedom and 7 the least amount of civil freedom. For example, in 2005 Freedom House calculated the Brazilian, Ecuadorian and Tanzanian civil liberties index as 3, while the civil liberties index of Zimbabwe corresponded to a rating of 6.

  It is important to mention the general acknowledgement that the Freedom House indices are sometimes accused of being biased, even if unintentionally, owing to their pro-North American perspective.

  Thus, according to the International Relations Center – Right Web[5], 'Although frequently cited in press reports and academic works, the reports and studies produced by Freedom House and its affiliates have been criticized for their alleged partiality towards US interests.'

2.5   Indicators of the social dialogue dimension

Anker *et al.* suggest that social dialogue 'refers to any type of negotiation, consultation or exchange of information between representatives of governments, employers and workers on issues of common interest relating directly to work and economic and social policies' (Anker *et al.*, 2002: 55–56). According to the ILO, 'a prerequisite for social dialogue is the existence of social partners that have both the capacity and the will to engage responsibly in the various forms of social dialogue at different levels' (ILO, 2003c: 70). The process of social dialogue can help to build efficient labour market institutions and to achieve consensus on issues concerning decent work deficits. The extent to which workers can express themselves on work-related matters and participate in defining their working conditions is essential: 'The ability of workers to organize freely to defend their interests collectively in negotiations with the employer is a pivotal element of democracy in the workplace and effectiveness of social dialogue' (Anker *et al*, 2002, 55–56). For the social dialogue dimension, we suggest the following indicators:

- Union density rate

  This refers to the number of union members expressed either as a percentage of the non-agricultural workforce or as a percentage of wage and salaried workers. Kuruvilla (2003: 5) notes that on some occasions union density has been utilized as an evaluation of union strength and the capability of social dialogue. He explains that in industrialized countries, a positive correlation between union density and any other measure of tripartite or bipartite industrial relations has been observed. Nevertheless, this measure is problematic in non-industrialized countries where this correlation is not straightforward. In other countries, union density as an illustration of union strength and as a base for social dialogue is dubious because unions are not independent and are subject to strict control. Another important point is the problem concerning the quality of data, particularly in terms of data compilation methodology and density estimation.

An additional difficulty of the union density rate noted by Kuruvilla (2003: 30) is that this indicator does not automatically describe the presence of real social dialogue. Nevertheless, Ghai (2002: 25) notes that 'in general, the higher the union density, the stronger the defence of workers' interests in negotiations with employers and the government, and the greater the participation by workers in matters affecting their work'.

- Collective bargaining coverage rate
  This refers to the number of workers covered by a collectively negotiated wage agreement. Kuruvilla (2003: 5) notes that collective bargaining coverage has been considered an evaluation of bipartite industrial relations in countries where negotiation was usually executed at the workplace. Collective bargaining coverage has also been considered a measure of tripartite industrial relations in countries where negotiation was attained at the national level. This is a concrete measure of the real degree of social dialogue because it gives a quantifiable indicator of the number of workers actually covered by collective bargaining agreements. However, in many countries collective bargaining coverage does not represent the right to bargain. Kuruvilla (2003: 31) notes that collective bargaining coverage (even if measuring the force of trade union activity in the workplace) does not estimate the quality of collective agreements.

The two social dialogue indicators of union density and collective bargaining coverage should be used carefully. Hepple (2003: 19–20) noted that in industrialized countries, union density and collective bargaining coverage have dramatically declined and that the employment contract has lost much of its analytical value because paid work is increasingly performed outside conventional employment relationships.

It is also important to note that in cases where workers are allowed to form and/or join organizations and to bargain collectively, there are often requirements and restrictions in national legislation which restrict their movements. In general, collective bargaining agreements (and that implies collective bargaining negotiations) are concluded in sectors and enterprises where there is a significant number of permanent workers. Thus, these agreements usually defend the rights and concerns of permanent workers, whereas those of daily, temporary or seasonal workers (typically the majority of workers in the construction sector) are often not considered.

- Strikes and lockouts
  Chernyshev (2003: 2) noted that one measure of the failure of social dialogue is the utilization of strikes or lockouts. As he observed, the ILO Resolution concerning statistics of strikes, lockouts and other action due to labour disputes gives the following definitions:

  - A strike is a temporary work stoppage effected by one or more groups of workers with a view to enforcing or resisting demands or expressing grievances, or supporting other workers in their demands or grievances.
  - A lockout is a total or partial temporary closure of one or more places of employment, or the hindering of the normal work activities of employees, by one or more employers with a view to enforcing or resisting demands or expressing grievances, or supporting other employers in their demands or grievances.

  Industrial action – strikes and lockouts – could consequently be the most important characteristic of social dialogue, in terms of media coverage and public impact. At the same time, in some cases the deficiency of strikes (and lockouts) can signify the absence of the right to strike (or lockout) and/or a fragile social dialogue. In order to be able to compare international measures of strikes and lockouts, Chernyshev presented the 'rate of days not worked per 1000 employees' as the most useful indicator. Usually, this threshold is determined in terms of the number of workers involved, the length of the dispute, the number of days lost, or a combination of all or some of these measures. Nevertheless, Kuruvilla (2003: 32) also notes that it is important to use these indicators carefully because the absence of strikes and lockouts will not necessarily mean a positive interpretation of social dialogue.

  According to the ILO (LABORSTA), in 2004, five strikes and lockouts were registered in Peru in the construction sector out of a total of 107 strikes and lockouts in the entire economy. During the same period, 11 strikes and lockouts were recorded in Algeria in the construction sector out of 35 for all sectors. Even if this data does not give any details about the exact number of days not worked, it does give an idea of the great number of strikes and lockouts in these countries.

- Degree of participation in decision-making
  Another feature of social dialogue suggested by Ghai (2002: 28) concerns workers' participation in the running of their enterprise, whether in the formal or informal economy. Workers' participation can

consist of a broad range of factors, varying from representation on the governing boards and executive committees to playing an active role in the management of training by occupational health and safety committees. The degree of participation in decision-making processes can be understood by studying labour laws, institutions, procedures and practices existing in each country.

- Participation at the national level
  In this case, we refer to direct participation in policy formulation and implementation using various measures, such as the representation of different social and economic groups on ministerial committees. According to Ghai (2002: 29), there are no simple indicators to measure the degree or efficiency of social dialogue at the national level. The degree of participation in decision-making suggests that it is important to analyze the laws, institutions, procedures and powers of consultative bodies and then their actual performance.

  It is also important to highlight the significant role of the state in facilitating and promoting all forms of social dialogue, 'creating the overall environment within which contributions of workers, employers and other concerned civil society groups are elicited and reflected in policy outcomes' (ILO, 2003c: 70).

2.6 Synthesis

This chapter has presented a broad set of decent work indicators based on the four main components of decent work. The primary objective has been to select indicators that would be easy to measure in all countries. However, we recommend that these indicators are adapted to the specific conditions and customs of each country, city and economic sector, in order to be more precise. Whenever possible, additional indicators of decent work should also be considered to give a broader view. Initially, the idea was to find a set of indicators with no ranking between them. Ideally, they should all have the same weighting as all the different indicators are measuring a precise characteristic of one of the four components of decent work. We acknowledge that in some situations, some indicators might be more complicated to calculate and monitor. For that reason, we have proposed to use qualitative information as well, in order to complete, and sometimes to confirm, the available quantitative data.

For the case studies presented in this book, we have suggested some indicators from the list in this chapter which are more pertinent for the construction sector. Thus, the selected indicators presented in Table 2.1 have also been classified and listed according to the four key components of decent work: employment, social security, workers rights and social dialogue.

Table 2.1   Indicators of decent work adapted to the construction sector

---

I Indicators of employment

1 Unemployment rates 1990–2000[1]
1.1 Unemployment rate in all sectors at the national level
1.2 Unemployment rate in all sectors at the city level
1.3 Unemployment rate in the construction sector at the national level
1.4 Unemployment rate in the construction sector at the city level

2 Low wage rates 1990–2000[2]
2.1 Low wage rate in all sectors at the national level
2.2 Low wage rate in all sectors at the city level
2.3 Low wage rate in the construction sector at the national level
2.4 Low wage rate in the construction sector at the city level

3 Safety at work 1990–2000[3]
3.1 Working days lost owing to accidents in the workplace per 100,000 wage earners in all sectors at the national level
3.2 Working days lost owing to accidents in the workplace per 100,000 wage earners in all sectors at the city level
3.3 Working days lost owing to accidents in the workplace per 100,000 wage earners in the construction sector at the national level
3.4 Working days lost owing to accidents in the workplace per 100,000 wage earners in the construction sector at the city level

4 Hours of work, 1990–2000[4]
4.1 Hours of work in all sectors at the national level
4.2 Hours of work in all sectors at the city level
4.3 Hours of work in the construction sector at the national level
4.4 Hours of work in the construction sector at the city level

5 Legislation on working conditions 1990–2000
Factual evidence on national legislation concerning paid holidays, hours of work per week, remuneration of overtime, night work, shift work and hygiene/comfort in the work place should be analyzed. The same information should be compiled for cities, because specific legislation could have been introduced by the local authority to deal with some or all of these working conditions. In addition, legislation which is specific to the construction sector should be analyzed for both the national and city levels.

6 Employment and informal sector 1990–2000
When data concerning the indicators of employment in the construction sector (unemployment rate, low wage rate, safety at work and hours of work) is available for the informal sector, it must be presented.

II Indicators of social security

7 Public social security coverage rate 1990–2000[5]
7.1 Social security coverage rate in all sectors at the national level
7.2 Social security coverage rate in all sectors at the city level
7.3 Social security coverage rate in the construction sector at the national level
7.4 Social security coverage rate in the construction sector at the city level

(Continued)

Table 2.1 (*Continued*)

8 Old age pensions 1990–2000[6]
8.1 Old age pension coverage rate in all sectors at the national level
8.2 Old age pension coverage rate in all sectors at the city level
8.3 Old age pension coverage rate in the construction sector at the national level
8.4 Old age pension coverage rate in the construction sector at the city level

9 Social protection for workers 1990–2000
Factual evidence about national legislation concerning the legal age of retirement and paid sick leave, in both the public and the private sectors in 1990 and 2000 should be analyzed.

It is also important to know whether there are workers' associations or cooperatives and NGOs that contribute in some way to promoting social security. The specific characteristics of the construction sector at both the national and city levels should be analyzed.

10 Social security and informal sector 1990–2000
When data concerning the indicators of social security in the construction sector (social security coverage rate and old age pensions) is available for the informal sector, it must be presented.

III Indicators of workers' rights

11 Wage inequality between genders 1990–2000[7]
11.1 Wage inequality between men and women in all sectors at the national level
11.2 Wage inequality between men and women in all sectors at the city level
11.3 Wage inequality between men and women in the construction sector at the national level
11.4 Wage inequality between men and women in the construction sector at the city level

12 Wage inequality according to workers' places of birth 1990–2000[8]
12.1 Wage inequality between natives and foreigners in all sectors at the national level
12.2 Wage inequality between natives and foreigners in all sectors at the city level
12.3 Wage inequality between natives and foreigners in the construction sector at the national level
12.4 Wage inequality between natives and foreigners in the construction sector at the city level

13 Child labour 1990–2000[9]
13.1 Child labour rate for all sectors at the national level
13.2 Child labour rate for all sectors at the city level

14 Legislation on workers' rights 1990–2000
Factual evidence on national legislation or legally binding conventions that outlaw discrimination against workers (owing to age, health status or nationality) should be stressed. The same information should be compiled for cities, because specific legislation could have been introduced by the local authority to deal with some or all of these workers rights. Legislation which is specific to the construction sector should also be analyzed for both the national and city levels.

15 Workers' rights and informal sector 1990–2000
When data concerning the indicators of workers' rights in the construction sector (wage inequality between genders, wage inequality according to workers' places of birth and child labour) is available for the informal sector, it must be presented.

Table 2.1 (*Continued*)

IV Indicators of social dialogue

16 Union density rates 1990–2000[10]
16.1 Union density rate in all sectors at the national level
16.2 Union density rate in all sectors at the city level
16.3 Union density rate in the construction sector at the national level
16.4 Union density rate in the construction sector at the city level

17 Collective bargaining coverage rates 1990–2000[11]
17.1 Collective bargaining coverage rate for all sectors at the national level
17.2 Collective bargaining coverage rate for all sectors at the city level
17.3 Collective bargaining coverage rate for the construction sector at the national level
17.4 Collective bargaining coverage rate for the construction sector at the city level

18 Legislation on social dialogue 1990–2000
Factual evidence on national legislation or any legal framework concerning social dialogue (union density rights, collective bargaining rights, right to strike, etc.) should be stressed. The same information should be compiled for cities, because specific legislation could have been introduced by the local authority to deal with some or all of these social dialogue rules. Legislation which is specific to the construction sector should also be analyzed for both the national and city levels.

19 Social dialogue and informal sector 1990–2000
When data concerning the indicators of social dialogue in the construction sector (union density rate, collective bargaining coverage and/or others) is available for the informal sector, it must be presented.

[1] Calculated by measuring the proportion of the working age population unable to find work.
[2] Measured by the number of employed persons earning less than half the median wage at the national level. For city levels, the low wage rate could also be based on the national level.
[3] It is recommended to use wage employment as the denominator in order to facilitate comparisons between sectors and between cities.
[4] Calculated by measuring the proportion of wage earners working more than 48 hours a week.
[5] Calculated by measuring the proportion of all employed persons of the working age population insured against the risks of unemployment and sick leave.
[6] Calculated by measuring the proportion of people aged 65 or over without pension coverage.
[7] Measured by the wage ratio between the median male wage and the median female wage.
[8] Measured by the wage ratio between the median wage for natives and the median wage for foreigners. Can be completed or replaced by the wage inequality between local and national workers, depending on the case analyzed, for all four indicators.
[9] Measured by the proportion of children between 10 and 14 not attending school. In this case, it may be difficult to find information concerning only the construction sector.
[10] Calculated by measuring the proportion of all employed persons (in the private and public sectors) of the working age population members of a trade union. In this case, if possible, data should be distinguished for private and public sectors.
[11] Calculated by measuring the proportion of all employed persons ((in the private and public sectors) of the working age population covered by a collectively negotiated wage agreement. In this case, if possible, data should be distinguished for private and public sectors.

Notes

Chapter opening image: Construction worker at a building site in Jordan. Photograph courtesy of J. Maillard, ILO.

(1)   'Employment opportunities' – opportunities for work; 'unacceptable work' and 'decent hours' – work in conditions of freedom; 'adequate earning and productive work' – productive work; 'fair treatment in employment', 'balancing work and family life', and 'social dialogue and workplace relations ' – equity and dignity at work; 'safe work environment', 'social protection', 'stability and security of work' – security at work.

(2)   There is no international definition of the working hours threshold, so this can be defined on a country by country basis.

(3)   http://laborsta.ilo.org

(4)   http://www.iess.gov.ec (consulted February 2009).

(5)   http://rightweb.irc-online.org

3 Local Authorities and the Construction Industry

Mariana Paredes Gil and Edmundo Werna

This chapter begins with a discussion on the concept of decent work in the context of urban areas in developing countries. Next, the chapter analyzes the evolution and new roles of local authorities in a globalizing economy during a period when decentralization has been promoted. The subsequent section analyzes the labour aspects of the construction industry. The chapter then brings together the themes raised in the previous sections by analyzing the role of local authorities in promoting decent work in the construction sector.

3.1   Decent work in urban areas

Cities and towns are not merely places to live, but also places to work. Given the magnitude and rapidity of urbanization, a crucial challenge is to create more and better quality employment in urban settlements. Work is mainly and often the only way to avoid urban poverty. Urban areas are places of opportunity and have provided jobs for enormous numbers of rural migrants and urban-born residents. However, in the developing world the expansion of urban populations is outpacing employment opportunities and urban areas need to be more efficient in creating decent jobs.

The ILO report *Global Employment Trends 2005* points out that there are 184 million people in the world who do not have jobs. This number increases to at least one billion if underemployed people are taken into account. Over the next ten years, the ILO estimates that 500 million people will join the world's job markets, most of them young people in developing countries. They will join the 184 million unemployed and the 550 million working poor, all wanting to use their talents and abilities in a productive and gainful way. Therefore, a large number of jobs must be provided by the end of this decade, simply to employ all those entering the labour market. This would require not only improved economic growth, but also policies and programs to increase the impact of this on opportunities for decent and productive work. The inability of cities and towns to productively absorb the influx of urban workers and generate enough quality jobs has led to rising levels of poverty and insecurity.

At the same time, cities and local governments have a number of comparatively advantageous areas for employment creation, although these are often misunderstood or poorly exploited. A large number of people have to resort to informal employment and many work in precarious conditions. Many in the informal economy are working long hours for low pay without any form of representation or social protection – often in dangerous, and sometimes violent and illegal activities.

Apart from a number of publications produced, notably by the ILO (which can be classified as 'grey literature'), publications on urban development scarcely analyze decent work in urban areas, and can be described as if 'urban analysts "have entered the city through the house and the bathroom", i.e. through housing and residential infrastructure, rather than through the place of work and the market' (Miller & Cohen, 2008: 4).

However, there are some notable exceptions, which will be analyzed below. In addition, a number of contributions approach aspects of decent work, although without reference to this broad concept. They will also be analyzed below. This section will show that the role of local authorities has been mentioned but has not received the in-depth analysis it deserves. Prominent exceptions include two articles by members of the research team which produced the present book (Lawrence *et al.*, 2008; Paredes Gil *et al.*, 2008).

3.1.1 Employment dimension

The following paragraphs present a general analysis of the key components for the employment dimension of decent work in the specific context of working conditions in urban areas.

Employment generation

By far the bulk of the urban literature dealing with labour issues concentrates only on employment generation. Many publications which analyze urban economy broadly have also brought to light the issue of employment generation across different sectors. Apart from these publications, traditionally there has been a focus on the informal sector, self-employment and on small-scale enterprises, with particular emphasis on the populations of low-income settlements. The construction industry is frequently mentioned for its employment-generating role (UNCHS-ILO, 1985; Werna, 2001). The ILO has contributed with a number of publications, including a comprehensive report (ILO, 1996) for the World Summit on Human Settlements held in Istanbul, 1996. Other publications referring to specific aspects of employment generation significant to urban areas include, for example, the informal sector (Musiolek, 2002) and community contract (Tournée & van Esch, 2001).

The literature in urban development has also often used the concept of livelihoods, which interfaces with employment generation (Rakodi & Lloyd-Jones, 2002). However, it is important to bear in mind that, 'livelihoods' and 'decent work' are not the same thing. Livelihoods include some aspects of employment creation but also issues related to other subjects, such as land and nutrition; in essence it has a *community-centred approach*. Decent work includes a comprehensive view of employment generation and is linked to the other aspects of the 'world of work'. It has a *labour-centred approach*. Nonetheless, the urban livelihoods literature has contributed to the understanding of income generation.

One important set of publications usually linked to livelihoods (although sometimes to income generation specifically), is those dealing with assets. The importance of assets has been widely promoted by the World Bank. Types of assets include human, social, physical, natural and financial capital. This analysis is understandably important from a decent work perspective, as it focuses on the types of capital necessary for a person to generate income. Apart from a number of authors (mostly related to or influenced by the World Bank) who have a more holistic approach to assets, contributors to urban development studies have usually paid attention to physical capital. For example, the role of housing as a basis for employment – such as home-based enterprises and the role of housing for rest and recuperation of workers, etc. (Werna, 2000). This echoes the earlier quotation from Miller & Cohen (2008: 4), that urban analysts 'have entered the city through the house and the bathroom'.

Literature on what can be regarded as physical capital continues to be produced, such as the in-depth analysis by Tipple (2005) about home-based enterprises in cities in the developing world. Majale (2008), in turn, provides a recent analysis from the perspective of decent work by linking employment creation to urban planning in Kenya.

Since the mid-1990s, there have been a growing number of urban studies dealing with social capital. Several studies have identified a link between social capital and economic growth in developing countries, including income generation. Krishna & Uphoff (1999), for example, found a positive relationship between levels of social capital (as measured by informal networks and mutual support) and the performance of settlements with a watershed conservation and development program in India. Narayan (1997) found a link between involvement in voluntary associations and household welfare in Tanzania. Bazan & Schmitz (1997) carried out an in-depth study in the town of Dois Irmaos in Southern Brazil which highlighted a strong association between the development of the local industrial community and their stock of social capital. Both UNV (2001) and Werna (2000) provide a broad analysis of the role of social capital in urban development, with implications for employment generation. A more recent contribution related to decent work is by Mitra (2008), who focuses on the importance of social capital for generating or improving employment based on data from New Delhi, India, with special attention to migrant workers and the informal sector. This contribution also includes an analysis of the construction industry and highlights its importance.

Two other recent publications on employment generation – beyond the assets approach in general and in the specific realm of social capital in particular – are Kantor (2008) and Miller & Cohen (2008). The first contribution focuses on income generation via employment diversification in Kabul, Afghanistan, and includes an analysis of social protection, another key component of decent work. Miller & Cohen (2008) wrote an ILO contribution to the Cities Alliance (an initiative of the World Bank, UN-Habitat and other partners). This text includes a broader rage of elements of employment generation following the decent work concept. Miller &

Cohen (2008) also emphasize the role of the construction industry as a generator of employment in urban areas. However, while local authorities have been mentioned in Miller & Cohen (2008) and indeed in other papers noted before, they did not receive attention as a distinct subject of analysis.

Contributions on urban development have placed an emphasis on the role employment generation plays in urban poverty reduction. However, while it is important to create jobs for newcomers or those who are unemployed, there are a large number of people already in the urban labour market experiencing other types of decent work deficits. This also has an impact on urban poverty. Therefore, it is also essential to address the other aspects of decent work.

Employment conditions

Inadequate working conditions still abound in urban areas. For instance, the ILO (2006e) notes that the figures for health problems and accidents across different sectors of urban work are still high, especially for the majority of developing countries. Many publications within the urban development literature mention the existence of poor working conditions in general, but without an in-depth analysis. One exception specifically related to health is the literature on healthy cities promoted by the World Health Organization and its corresponding Health Cities project. Health Cities includes a perspective on 'health in the workplace' which overlaps with the occupational health and safety perspective but also includes a settings approach. This has helped to highlight aspects of the employment conditions of urban workers (Barten *et al.*, 2006; Werna *et al.*, 1998).

Recently, Barten *et al.* (2008) provided a broad review of occupational health and safety in urban areas, with particular attention to informal workers. While the ILO has produced many publications on occupational health and safety, as well as other aspects of employment conditions, the urban context has not gained attention as a predominant locality of analysis. There are a few urban-related papers providing data on specific places.

3.1.2 Social protection

Today, many urban workers do not have access to an adequate system of social security, including health care, pay for holidays and protection against loss of pay when they are unable to work due to unemployment, ill-health, accidents or old age. Lack of social protection is a major cause of urban poverty, especially – but not only – for those who work in the informal sector. Living and working conditions expose workers to risks on a daily basis and these go beyond occupational health and safety. Risks of illness, disability, accidents and premature death are high when there is no clean water or proper sanitation, when there is exposure to fire and flood, dangerous electrical wiring, casual use of toxic substances, dehumanizing working conditions, overcrowding, crime and substance abuse.

Exposure to these multiple risks is high and people living and working in low-income settlements are the least protected. The quality of health care is lowest in the areas where they live and work. Emergency services, such as fire brigades, are virtually absent in these settlements. Awareness campaigns and subsidies for safety measures often do not reach the working poor. Therefore, the poor work hard to survive, but without social protection. Should one income earner in the household be injured or become sick, the household can fall into absolute poverty, child labour or debt bondage.

In response to their vulnerability, poor people in cities have at times mobilized their resources and organized their own risk defence through mutual health protection or community surveillance. The coverage and benefits of these schemes remain limited by the poverty of their members. Also, these are proving woefully unequal to the challenge raised by the HIV/AIDS epidemic (which could be an occupational health issue but in many cases is not related to the workplace). HIV/AIDS threatens the livelihood of many workers and those families, communities and enterprises that depend on them.

The urban development literature has generally neglected the above issues, perhaps with the exception of some analysis of the HIV/AIDS situation in urban areas. The ILO has produced a number of publications which provide data on social protection in urban areas (ILO, 2008a; 2008b). A contribution by Frota (2008) provides a broad coverage of the topic and explains the importance of decent work in combating urban poverty and exclusion.

3.1.3   Workers' rights

Informal workers in urban areas are particularly at risk because their relation with employers as well as the government is not regulated. Therefore, it is difficult or impossible for them to demand rights. Migrant workers are also often at risk because they have to get by with unacceptable working conditions in order to survive in a foreign city. Child labour is also a harsh reality which deserves special attention because of the moral imperative and, in addition, because it impacts on future adults. Today, with rapid urbanization, rising poverty and a growing number of children orphaned by the HIV/AIDS epidemic, young people are increasingly vulnerable to exploitation in illegal, underground and hazardous activities. In addition to these direct impacts, child labour undermines education, which is a prerequisite for children and youth to find decent and productive work later in life. The ILO estimated that in 2000 there were approximately 186 million child workers under the age of 15, with about 110 million under the age of 12[1].

Children working in cities tend to come from poor families. They work mainly in manufacturing, trade and domestic services. In all three sectors, they work long hours for low wages. Where traditional social regulation (apprenticeships) is not in operation, they are completely without protection. On the streets or on waste dumps, which are home and workplace for so many, they can be seen sorting rubbish, carrying loads and surviving any way they can. Girls, who are mainly in domestic service, are exposed to physical, psychological or sexual abuse from their employers.

As already mentioned, the informal sector has been a predominant object of analysis in the literature which deals with employment generation in urban areas. A number of publications have also highlighted the plight of informal workers in relation to their rights. However, there is no consensus in sight regarding a comprehensive solution. Recommendations range from those who advocate formalization of the informal sector to those who advocate that it should simply not be discriminated against; that if 'left alone', it will thrive. One way or another, there is a message regarding the need to respect the rights of informal workers.

In parallel, the literature on gender and (urban) development, which has grown steadily since the 1980s, has repeatedly analyzed issues relating to deficits in the rights of female workers. It has highlighted the fact that women are worse off than men and that this type of gender inequality needs to be addressed. A number of recent papers highlight the plight of women in the urban labour market, for example Mitra (2008), Majale (2008) and Kantor (2008) among others.

Working children have also been the subject of much attention, being one aspect of a broad analysis of the plight of children in cities and towns. This analysis also included other issues, such as housing and homelessness, health, violence and education, etc. (Christensen & O'Brien, 2003; Werna *et al.*, 1999). Some contributions that address working children from an urban labour perspective come from Barten *et al.* (2008), Bourdillon (2006), Kantor (2008) and Srinivas (2008). There seems to be some consensus that working children are a very serious problem, in need of being addressed. The role of many actors as part of the solution has been mentioned – from government (both central and local) to NGOs, private sector actors, communities, religious institutions and others. A definite solution does need a concerted effort owing to the complex set of causes that generate this social problem.

There has been less attention related to urban migrants, although they constitute a large and particularly important group of workers; their key issues concern rights and working conditions. There are many papers on migrant workers in general but these lack an urban perspective. Others have a sector-based focus – including the construction sector – but again, without an urban perspective. One notable exception is Klink (2008), who notes that the literature on the relationship between the international migration of workers and development lacks a perspective specific to urban areas. His contribution analyzes this issue and makes recommendations for a research agenda. Other papers include references to migrant workers in urban areas using a broader analysis, such as Srinivas (2008).

3.1.4 Social dialogue

The low level of organization among the majority of urban workers – especially, although not only, in the informal economy – is a major cause of concern. Socially viable cities cannot exist without the involvement and fair representation of the majority of urban workers in decisions that affect them.

Social dialogue includes participation. However, the bulk of the literature on participation in urban areas focuses on low-income communities. It dates back to the 1960s with the pioneering work of John Turner (Turner, 1967; 1968; 1977; Turner & Fichter, 1972). Today, this subject is represented by a large volume of publications (Viloria-Williams, 2006; World Bank, 2005b). It is a consolidated approach which has been, and continues to be, widely used by both local and international actors.

Participatory approaches have brought benefits for low-income communities, leading to the physical upgrading of their settlements as well as to the improvement of social services, including educational and health facilities. They have also helped local authorities and other stakeholders to better understand the needs of low-income communities. However, these approaches have been limited in the promotion of decent work.

Since the 1990s, the importance of a city-wide participatory approach has been acknowledged. For example, the concept of 'participatory budgets', pioneered by the city of Porto Alegre in Brazil, has been debated around the World. This approach is applied in many municipalities in different (developing and also some developed) countries. It is promoted by international organizations such as UN-Habitat. However, this kind of approach has been limited in relation to the promotion of decent work issues.

Both community-based and city-wide participatory approaches have brought some labour-related benefits, especially regarding employment generation in the context of small enterprises and self-employed producers. While this is commendable, there are many other issues which need to be addressed in order to achieve decent work in urban areas. For example, a large number of workers still lack the basic elements of social protection and/or respect of their rights. These are important issues for reducing urban poverty, yet are seldom addressed in participatory processes.

Two recent papers have attempted to address this gap, using a broad view on city-wide social dialogue (Van Empel, 2008; Van Empel & Werna,

2008). While different forms of social dialogue have been extensively used by the ILO and other institutions to promote decent work, these contributions explain that city-wide social dialogue is still developing. They argue that social dialogue has a potential to benefit different stakeholders, with some illustrations from different places around the world.

Williams (2007) wrote a paper which provides a complementary perspective to the above. It focuses on how the ILO official constituents – associations of unionized workers and formal enterprises, as well as central governments – could engage in a participatory process together with informal workers and low-income communities. Other actors, such as cooperatives and local governments, are also included in the picture. Again, the construction industry is mentioned for its employment generating potential. However, it was beyond the scope of the paper to provide an in-depth analysis of the role of local authorities.

Another important paper, by Jason (2008), focuses on the process of social dialogue in Dar es Salaam, Tanzania. Jason presents a concrete case study of how the informal construction workers organized themselves and have benefited from such a process.

All the abovementioned types of social dialogue provide important participatory platforms for improving the generation of employment and quality of work in urban areas.

3.1.5 *Cross-cutting analyses*

There are two key ILO papers which present a broad view of decent work in urban areas (ILO, 2004a; ILO 2006e). The first provides useful data in different aspects of decent work and highlights some actions taken to address the existing deficits. The second is a recent summarized presentation of an ILO strategy in urban areas.

With reference to the above papers, Werna (2008) presents decent work in his introduction to a special issue of the journal *Habitat International*, which focuses on urban labour. Most of the papers included in this special issue address specific aspects of decent work and have been mentioned before in this chapter. Three papers with a broader perspective are Srinivas (2008), Lawrence *et al.* (2008) and Paredes Gil *et al.* (2008). Srinivas links the international scene (e.g. globalization and international trade) to the local economy, dove-tailing into concrete illustrations from one country (India), one city (Bangalore) and one sector of the urban economy (construction). The joint paper by Lawrence *et al.* (2008) focuses on indicators of decent work, whilst Paredes Gil *et al.* (2008) brings the discussion to the specific context of Santo André (Brazil).

The literature in this chapter includes many accounts of decent work deficits. Hence, a fundamental question is why urban workers face so many problems. As Werna (2008) notes, there is still probably not enough attention being paid to urban labour. This is reinforced by the traditional,

but still current, notion that the answer lies in rural economic development, which per se would bring the urban poor back to rural areas and
address urban poverty. This proposal may not be shared by urban planners, but still prevails among professionals in many other fields that do
influence what happens in cities and towns. Many also share the idea
that it is through policies of the Ministry of Labour at the national level
that change can occur. This may be true, but only partially. Of course not
all problems can be addressed by urban policies. For example, much depends also on economic growth and a stable political environment in each
country. Nonetheless, policy decision-makers with a stake in urban development can and should make a difference. Many papers presented above
highlight the role of local authorities. Yet such authorities still deserve a
specific focus of attention.

3.2   Local authorities

Decentralization has become a major concept and trend throughout the
world, especially during the 1990s, with significant implications for local
authorities. In particular, local authorities play a strong role in the construction sector and related services, either via the direct execution of public works and/or in some form of partnership with the private sector. At
the same time, as already noted, decent work continues to be a crucial issue, particularly in developing and transition countries.

   This section of the chapter is based on an analysis prepared by Klink
(2006) as a foundation for the research which generated the present book.
In short, Klink noted that, although a lot of theoretical and empirical work
has been undertaken, the role of local authorities in generating decent
work has been largely ignored.

3.2.1   Evolution of the role of cities in the global economy

An important trend observed in recent decades is the new territorial and
competitive role of cities in the global economy: local stakeholders, instead
of passively depending on macro-economic and micro-economic forces in
movements related to globalization, have become increasingly concerned
about the potential competitive advantages of large cities and metropolitan regions. This trend has grown in parallel to changes in overall macro-
and micro-economic frameworks in Europe and the USA since the 1970s,
and in many developing countries since the mid-1980s. As a result of this
economic restructuring, cities and metropolitan areas have been increasingly engaged in new challenges related to local development and income
and employment generation without depending exclusively on national
initiatives.

Since the 1970s, national governments have increasingly retreated from traditional Keynesian-style, active macro-economic management policies aimed at full employment and income generation. Therefore, these policies have become difficult to implement. They have lost some of their effectiveness in the context of an increasingly deregulated international economy characterized by massive volatile flows of international financial capital. At the same time, the continuing tendency towards de-regularization and trade-liberalization has had impacts on the behaviour of companies, especially of those that have been operating within relatively protected domestic markets. Hence, these companies initiated a series of micro-economic adjustments aimed at managerial and technological modernization. These processes increased all levels of productivity, but they did not always create immediate positive effects on employment. Indeed, in many cities distinguished by relatively obsolete industrial structures, the increase in productivity and the shift away from industrial towards tertiary employment resulted in losses in the total number of those working in formal employment.

A great number of cities and metropolitan areas have been affected by the impact of macro- and micro-economic restructuring. The bottom line of experiences in different cities was to create the right conditions for internal development, taking advantages of the local skills available in each city. The essence of these experiences was the awareness that a productive mobilization of public and private actors and skills would permit the creative use of globalization, instead of becoming its passive victim. Simultaneously, it could also improve urban productivity and citizens' salaries, working conditions and quality of life.

A second point, underlined by Klink (2006), regarding the new role of cities and urban regions is related to their potential to encourage cooperation among local stakeholders through participatory processes. Klink notes that the nation state has been challenged: on the one hand, its macro-economic apparatus has lost effectiveness in light of the size and instability of massive flows of financial capital at the global level. On the other hand, given the global transformation towards more democratic and diversified local communities, national authorities have lost out against local and metropolitan systems of governance that are closer to local constituencies. Considering the increase of the network society, local communities in urban areas would also have new ways and means to evade conventional national borders and networks. Today, they can exchange experiences on a global level.

In the specific case of developing countries, a gradual process towards decentralization has been observed. Additionally, an increasing number of cities and metropolitan regions in developing countries are experimenting with innovative tools of direct democracy, such as participatory budgeting, city visioning, cooperatives and strategic planning.

3.2.2 New roles for local authorities

As a consequence of ongoing changes to the role of cities within the national and international development context, the roles of local authorities and urban management are also changing. Klink highlights changes
in the culture of urban management and the role of local authorities in the
European and North American context since the 1970s. Three aspects are
relevant.

First, local leaders are starting to incorporate new themes and change
previously established priorities on policy agendas. Therefore, while the
management of a set of urban services (such as housing, basic sanitation,
health and education) does not disappear from the policy agenda, at the
same time, issues like competitiveness, sustainable employment and income generation have gained strategic importance as a consequence of the
national government's gradual retreat from these areas.

Second, there is an evolving international situation characterized by an
increasing level of competition among cities which constrains local authorities to engage in innovative and area-based strategies for urban and
economic revitalization. An increasing number of cities are beginning to
adopt more flexible institutional arrangements and are willing to acquire
new responsibilities. Klink (2006) underlines that there are many examples of this trend in Latin American countries, particularly where there are
strong local governments.

Third, there is also a growing perception of new urban governance.
Since the 1990s, local governments have launched democratic processes
that enable many actors to participate in a multi-stakeholder society. This
trend, noted by Helmsing (2001: 9–10), symbolizes a contrast with the previous custom, by which local authorities were considered as mere implementing agents, frequently without much participation from public and
private stakeholders. The concept of governance involves negotiation processes through which a set of governmental and non-governmental actors (such as community based associations, public-private partnerships,
labour unions, enterprises, etc.) work together for collective goods and
policy-making. Additionally, the result of urban planning itself is shifting
from comprehensive and detailed master plans towards more operational
and area-based strategic plans elaborated and discussed with a wide range
of stakeholders from the local community.

The ongoing shift in the culture of urban management and local governance has also become increasingly relevant for developing countries since
the beginning of the 1990s, especially in Latin America. Thus, a growing
number of cities are acquiring new responsibilities without formally being
directly responsible for these domains. Having recognized the impossibil-
ity of acting as single service provider, many local authorities are accepting
a 'new participatory framework that increasingly involves local actors in
delivering services and building physical infrastructure' (ILO, 2004a: 21).
With this new approach, local authorities have maintained a crucial role,

particularly in the establishment and implementation of a regulatory framework. For example, local authorities can define standards for infrastructure and construction, contracting processes and land regulations, as well as imposition for the private sector. Local authorities can also play an important role in articulating the needs of their cities and residents because they serve as their representatives at the national and international levels. Thus 'in this capacity, they need better access to policy-formulation at the higher level; they need to be involved in drawing up policies that will eventually affect them and the people they govern'. In addition, 'local authorities need to provide an enabling environment for local development, and in particular, to establish partnerships for employment creation' (ILO, 2004a: 21).

Klink (2006) notes that his review of the existing literature on the changing roles of cities and local authorities within national and international contexts enables some preliminary conclusions. In general, there is a lack of detailed studies on a decent work perspective within the traditional programs of local authorities in public works and other construction-related issues. However, ample references exist concerning the evolution of the role of local authorities within a globalizing economy. As a result, there have been original evaluations of local economic development and employment strategies in general (such as cooperatives, clusters, incubators and micro-finance, etc.). However, even within these evaluations, there is a lack of a decent work perspective concerning the role of local authorities and their tasks with respect to construction and related services.

Another finding of the literature review is the existence of empirical research concerning the role of local authorities as important actors in poverty alleviation strategies. Klink (2006) found that in Latin American countries that have implemented a strategic decentralized policy, local governments are concerned about establishing and implementing inclusive policies through safety nets, minimum income programs and other compensation plans targeted at the most vulnerable population groups.

In sum, this section has introduced the specific analysis of local authorities, which will now be complemented with an analysis of their role in the construction sector. However, it is important to provide an overview of the construction sector itself. This book also includes information on related services (utilities) for the reasons explained in the introduction. However, as the focus is mainly on the construction sector, the next section does not include an analysis of utilities. The subsequent section deals with both construction and utilities through the perspective of the role of local authorities.

3.3   The construction sector

In 2001, the ILO published a report on labour in construction. It was the basic document for the tripartite international meeting *The Construction*

Industry in the Twenty-first Century: Its Image, Employment Prospects and Skill Requirements, which involved representatives of workers and employers, as well as national governments. The report was a major undertaking, prepared by Jill Wells with contributions from many leading authors from around the globe (ILO, 2001c). Follow-up work carried out by the ILO on working conditions in the construction sector confirms that this report remains pertinent. For these reasons, the present section is based on data from this report.

3.3.1   The construction sector: definition and general characteristics

The construction sector produces a wide range of products, from individual houses to large buildings and major infrastructure, including roads, power plants and petrochemical complexes. In most countries output is roughly equally divided between housing, other buildings and civil engineering projects. Although attention is mostly focused on new construction, the renovation and maintenance of existing structures accounts for almost 50% of total construction output in some of the more developed economies and an even greater share of employment.

The enterprises engaged in construction activity are equally diverse. They range from self-employed individuals providing a service to private house owners in the local community to multinational firms operating on a global scale. However, the vast majority of enterprises involved in on-site construction are small and local. Despite much talk of 'globalization' and the existence of an international construction industry, more than 95% of construction activity is still undertaken by firms from within the country, region or neighbourhood.

In recent decades there has been a marked shift towards flexibility. There is an increasing tendency among enterprises in construction (as in other industries) to outsource the supply of goods and services required in the production process. Building materials, plant and equipment are generally purchased or hired from other enterprises. Specialized services are supplied by subcontractors, and labour by 'labour agents'. Design and engineering services are also provided by quite separate professional entities.

Wells (1986: 56) provided a list of five indicators by which the actual performance of the construction sector may be evaluated in any country:

(1) The extent to which the construction projects in a country's development plan are actually implemented in the specified time.
(2) The percentage of imports in the total construction output.
(3) The degree of development of local skills and of local participation in contracting.
(4) The extent of development of local building materials industries.

(5) The overall efficiency/productivity of the construction sector and the extent to which construction plans are implemented within the cost limits set.

Using these five indicators, it is possible to evaluate the relative weight and performance of the construction industry in a national economy, but not in terms of employment significance. This book includes a proposal for specific indicators on decent work in construction. The next section presents information on the labour characteristics of the sector.

3.3.2   Implications for decent work

The following sections explicitly address the implications of the contextual conditions of employment in the construction sector on the promotion of decent work using the diverse roles and responsibilities of local authorities.

Employment generation and conditions of work

The products of the construction industry are fixed in space, so production takes place on a project-by-project basis with the production site constantly moving. This implies that the labour force should also be mobile. The construction industry has a long tradition of employing migrant labour. During the process of economic development, work in construction provides a traditional point of entry to the labour force for migrant workers from the countryside. Construction work is often the only significant alternative to farm labour for those without any particular skill or education, and it has special importance for the landless.

When the pool of surplus labour from the countryside has been used, or there is a shortage of local labour for other reasons, labour may be recruited from overseas. Migrant construction workers are generally from less developed countries and lower wage economies with labour surpluses. Many European countries also rely heavily on migrant workers to fill jobs in the construction sector. Workers are from poorer countries in Europe or further afield (Turkey or Africa). Migrant labour is also significant in the countries of the Arabian Gulf, with small populations and large construction programs financed by revenue from petroleum markets. In the past decade, migration for work in construction has become a significant phenomenon in East Asia, where huge differentials in demographic characteristics and wages have led to a 'siphoning' of migrant workers from lower to higher wage economies.

Despite migration flows to richer countries, the majority of construction workers are still in developing countries. The distribution of construction employment is almost the exact reverse of that of output. While three-quarters of output is in developed countries, three-quarters of employment is in the developing world. However, as many construction workers in these countries are informally employed, and therefore not counted in official data, the real number is much higher.

The reason for the greater employment-generating potential of construction activity in developing countries can be traced to differences in technology. There is a very wide choice of technology available for most types of construction and the technology adopted tends to reflect the relative cost of labour and capital. In the richer countries, where labour is expensive, machines have largely replaced workers in many of the tasks involved in new construction (although repair and maintenance is still very labour intensive). In developing countries, where labour is cheap, the majority of tasks are undertaken by manual methods with minimal use of machinery and equipment.

While construction has a large potential to generate jobs, instability of employment is one of the major problems the industry is facing. Fluctuations in demand, the project base of construction and the widespread use of the contracting system all conspire to make it difficult for contractors to obtain a steady flow of work which would allow them to provide continuity of employment. Hence, there is a constant friction between employers' needs for 'flexibility' and workers' needs for stable employment. It has become the norm for construction workers to be employed on a short-term basis, for the whole or part of a project. This implies no guarantee of future work. The number of casual and informal workers has greatly increased in both developing and developed countries.

In 'triangular employment relationships' (contractors – subcontractors and labour agents – workers), workers' rights are often unclear. Subcontracted workers may face less protection from the law than those who are directly employed. The same applies for the informal sector. This custom also imposes a considerable barrier for training in the industry. Spain is one good illustration because there has been a narrowing of the development of skilled labour there in recent decades. The heightened division of labour into ever more specialized trades, which is implicit in subcontracting, limits the range of skills that can be acquired in any one enterprise. This means that all-round craftsmen and general supervisory workers are very difficult to train. In many countries, the public sector used to provide stable employment and good ground for training, but its training role has diminished as public sector units have been disbanded.

Construction work is one of the most dangerous occupations in terms of health and safety. Data from a number of industrialized countries shows that construction workers are three to four times more likely than other workers to die from accidents at work. Many more suffer and die from occupational diseases arising from past exposure to dangerous substances, such as asbestos. In the developing world, the risks associated with construction work are much greater; available data would suggest three to six times greater. It is estimated that 95% of serious accidents involve workers employed by subcontractors. Most of these workers are on temporary contracts, which in a context of fluctuating demand, encourages them to work long hours in order to make the most of work while it lasts. They are also less likely than workers on permanent contracts to gain the training and experience required to work safely in a dangerous working environment and are in a weaker position to refuse to work in unsafe conditions. A construction worker with a fixed-term contract is three times more likely to suffer an occupational accident than one with a permanent contract. Informal workers are also particularly vulnerable. While many formal workers still lack training in health and safety, this is even more prominent in the informal sector. However, the causes of health problems and accidents are well known, and almost all are preventable.

Social security

The increased recourse to casual and temporary work, as well as subcontracting all around the world in the past 30 years, has resulted in a significant reduction in the number of construction workers covered by social security schemes:

> In some countries this has happened because temporary workers have been excluded from the provisions of labour legislation. In many more cases there is provision for temporary workers to receive benefits, but they are not claimed. There are also reports from a number of countries of total abuse, such as employers deducting contributions from wages but failing to forward them.
>
> (ILO, 2001c: 35).

In general, 'temporary employment status means that the majority of construction workers enjoy little or no social protection (income security or social security)' (ILO, 2001c: 43).

There is evidence from many countries that employers do not pay into social security funds for workers who are on temporary contracts. Hence, the workers who are most in need receive no health care, no holiday pay and no protection against loss of pay when they are unable to work due to unemployment, ill health, accidents or old age.

Rights at work

Basic labour rights are frequently ignored in the construction industry. 'In many countries construction workers are excluded by law from joining trade unions because of their temporary employment status, because they are self-employed or because they are foreign' (ILO, 2001c: 43). Discrimination between workers on the basis of gender, place of birth and temporary or permanent contracts in the terms and conditions of employment is common in both developing and developed countries. In regard to migrants specifically, while many countries have recognized their dependence on such workers and have attempted to regularize and control the process of migration for work in the construction industry, there are still many foreigners working illegally. These workers are extremely vulnerable to exploitation.

Social dialogue

Social dialogue with employers – and also with governments – has traditionally been a powerful mechanism for workers to bargain collectively for better wages and better working conditions. However, nowadays the vast numbers of temporary, casual, informal and unemployed workers find it very difficult to organize themselves and engage in social dialogue. Even worse, in some cases, the law does not allow temporary, self-employed or foreign workers to join trade unions. In general, in a large number of countries, 'both workers' and employers' organizations have been seriously weakened by the increased fragmentation of the industry. Collective bargaining has been undermined almost everywhere and collective agreements, where they exist, are applied to a small and decreasing proportion of the workforce' (ILO, 2001c: 30).

3.3.3   Recommendations for action

Given that the characteristics and trends of the construction industry that have generated the challenges and problems discussed in this chapter are likely to continue, it is important to discuss ways in which to address these challenges and problems. The following suggestions are also based on ILO (2001c).

There is evidence that the employment-generating potential of construction investment for some types of infrastructure may not have been fully realized. This stems from constraints in the planning and procurement of projects, as well as in the lack of capacity in the local construction industry, particularly in developing countries. In many of these countries, construction investment is at a very low level. The way forward is to expand the volume of output and employment in the sector, for example through the development of public-private partnerships and an appropriate choice of technology.

Innovative solutions should be considered regarding the training of workers. For instance, the training of master craftsmen to train young workers will improve the on-the-job apprenticeship system, as already shown in several countries. Another worthwhile way to supplement skills acquired through the apprenticeship system is to issue target groups with training vouchers, which they can spend according to their requirements.

This has been demonstrated, for example, in a project in Kenya. Also, the involvement of subcontractors, labour contractors and intermediaries in joint training schemes, with cost reimbursement, seems to be essential if these schemes are to be effective in meeting the real needs of the industry.

In terms of health and safety, in some countries it is still necessary to update the laws regulating this area. In many other countries, although appropriate laws are already in place, there is a problem of enforcement. By and large, there are never enough inspectors to police even the big sites, let alone the myriad of small ones. Corruption is also a problem in many places. The way forward is to change the role of labour inspectors to one of education and prevention, as opposed to inspection and prosecution. This is already being adopted in a number of countries, but should also be considered in several countries still lagging behind.

There have also been some experiments with safety cards. For example, the Construction Industry Development Board of Malaysia has pioneered a scheme to make every construction worker undergo a health and safety induction course. Each participant is issued with a green card. Those without the card would be barred from entering construction sites. Contractors who fail to send their workers to undergo the training have been threatened with blacklisting. Contractors for major projects are also required to send their management staff to attend the training course. A similar scheme is in operation in a number of developed countries, including Australia, Ireland and the United States. However, securing real improvement in occupational health and safety requires more than advice and training. To meet this challenge, many insurance initiatives have been applied in developed countries, which offer financial incentives to encourage employers to implement accident prevention strategies. For example, a lower annual premium if claims are reduced or a surcharge on excessively high levels of claims. Switzerland and Germany are examples of countries which use this kind of scheme, with significant advice and support offered to employers from the insurance providers.

An alternative that might be more appropriate in developing countries – where insurance schemes are not well developed – is to take the costs of health and safety measures out of competition by including them in the prime costs of a competitively tendered contract. Migrant workers comprise a significant share of construction workers in many countries. Therefore, they deserve special attention for inclusion in benefit schemes.

In terms of social security, where there are state insurance schemes that apply to permanent workers, attempts can be made to extend these to all workers. In many countries, however, a new approach may be required, with schemes specifically tailored to the needs of construction workers. In Australia, unions in the construction industry have responded to the prevalence of short-term employment by developing collective industry agreements at the state level for portable benefit systems. These schemes allow construction workers to accrue benefits on the basis of length of service in the industry, rather than with a specific firm. The Government of the Republic of Korea, in turn, has recognized the special needs of construction workers, introducing a law which is a mutual aid project for retirement allowances. Some states in India have been operating a Construction Workers Welfare Board which is funded by a levy on all building works.

In terms of social dialogue, it is important to seek new roles for trade unions and other actors. In many countries, particular sections of the workforce are not allowed to organize themselves and therefore cannot join trade unions. This is often the case with migrant workers. When this happens, the existing trade unions should campaign to remove the legal restrictions, therefore allowing all workers to join the unions. It is also important for trade unions to secure positive improvements in collaboration with employers. In Canada, for example, there have been joint activities to raise the levels of safety, quality and productivity.

While trade unions are adopting new roles, other new organizations are joining in campaigning for workers. In India, for example, the National Campaign Committee on Central Legislation for Construction Workers has campaigned to procure better legislation to protect workers in the sector. There are also cases of informal construction workers who have organized themselves, such as in Dar es Salaam, Tanzania.

All these suggestions may help to improve labour conditions in the construction sector. However, not all labour problems in construction stem from the characteristics of the industry. For example, much also depends on economic growth and a stable political environment in each country. But many of the suggestions in this chapter are now within the legal reach of local governments. In addition, local governments may also promote other suggestions through their municipal programs and related activities. Whether and how this has taken place so far is analyzed in the next section of this chapter.

3.4   Local authorities and decent work in the construction sector and related services

This final section brings together local authorities and the construction industry, adding related services to the analysis. This section is also based on an analysis prepared by Klink (2006) as a foundation for the research which generated the present book.

Local authorities have increasingly assumed new roles beyond the mere management of urban services (such as housing, basic sanitation, education and health services). The entrepreneurial dimension of public urban management (as reflected in the rise of strategies and employment generation, local economic development and poverty alleviation) has become a prominent feature of the new urban agenda with local authorities.

Nevertheless, there seems to be little evidence of how, at least at the local level, these managerial and entrepreneurial agendas reinforce and consolidate decent work programs, especially in the construction sector and related services.

The construction sector and related services are considered pertinent for the promotion of employment. In principle, the construction industry and related services constitute one of the major providers – and, in many instances, the primary provider – of work, particularly for unskilled or poor workers. In addition to direct employment on construction sites, the industry provides a large number of other jobs, such as in the production of building materials and equipment and post-construction maintenance. It is precisely for these reasons that it is important to analyze the construction sector and related services, and to evaluate the existing functions and responsibilities of local authorities in the promotion of decent work.

According to Radwan (1997), local authorities can play an important role in the promotion of employment and decent work in the construction sector and related services. First, local authorities can influence investment policies – especially in infrastructure and housing – in order to have a larger impact on employment creation and poverty alleviation: 'the investments that are of greatest benefit to the urban poor are all conducive to labour-intensive technologies – and hence job creation – and provide direct improvements to the urban environment' (Radwan, 1997: 323–324).

Second, the role of the urban informal sector should not be underestimated; local authorities should support it. Radwan stated that 'improving basic urban services and infrastructure provides one 'win-win' scenario that can improve incomes, productivity and working conditions in the urban informal sector' (Radwan, 1997: 324). Alliances between labour unions and informal sector associations should be encouraged in order to better organize and represent all types of urban workers. For this reason, Radwan (1997: 324) notes that there is a need for programs capable of providing 'basic social security, health care and urban services, [and which would] upgrade the physical working and living environment of the informal sector, improve productivity and incomes, and help the self-employed to organize and strengthen their bargaining power and be aware of their rights'. Decent work can also be promoted at the urban informal sector level under these conditions.

A third aspect is the development of alliances at the local level for greater employment and the promotion of decent work. Radwan (1997) notes that local governments 'have to increasingly forge partnerships with the private sector, employers and workers' organizations, different levels of government, and nongovernmental and community based organizations' (Radwan, 1997: 325).

The following paragraphs present some existing approaches that promote decent work using interactions between local governments and other relevant actors as participants in the promotion of employment policies. The first approach concerns partnerships between local actors and the private sector. Through privatization and changes in their procurement regulations, local authorities provide new market opportunities for the whole private sector, including local business. Nevertheless, local authorities have to be careful and implement appropriate policies, regulations and incentives in order to ensure the effective promotion of employment respecting the rights of workers and conditions in the workplace, as well as the other dimensions of decent work. The ILO (2004a: 24) noted that care should be taken to guarantee that this does not enhance the vulnerability of workers or working conditions. In some cases, privatization has been used by local authorities as a way to reduce costs. However, sometimes 'workers engaged ... by private companies, may be subject to poor working conditions, long hours and sub-standard wages; they may have no proper contract of employment and no social security benefits.'

The second approach is known as community contracting. One example of this approach is briefly presented in Box 3.1. This example of the Kalerwe experience shows how a local authority can provide technical support and create an enabling environment for community contracting in urban infrastructure. In general, local authorities can help communities to organize themselves and to become actors in employment promotion, often in infrastructure works. As noted by ASIST-Africa, 'the main aim of community contracts is to actively involve communities in the planning and implementation of construction activities. In a community contract, the community or community groups are always the implementer or contractor' (Tournée & van Esch, 2001: 34). According to the ILO, 'community contracts have been used for urban infrastructure construction and maintenance and for the delivery of public services' (ILO, 2004a: 26). Nonetheless, 'community contracts need a conducive environment and it is therefore important to enhance the capacity of local authorities to anticipate needs, design works, manage contracts and work in partnership with community groups' (ILO, 2004a: 37).

Box 3.1   Kalerwe: a community contract experience, Uganda

Kalerwe is a low-lying unplanned settlement in Kampala, Uganda, where the inadequate drainage network resulted in severe flooding during the rainy seasons. In addition to damage to property, living conditions became very unhygienic due to the overflow of latrines and the mixing of water with uncollected rubbish. The existing primary drain and four secondary drains were not functional. This situation provided a breeding ground for mosquitoes and the site was used as dumping ground for all kinds of waste, thus leading to an increased incidence of water-borne diseases.

Resident Committees (RC) and the Kampala City Council (KCC) were well aware that the solution to the flood problem would be the construction of a main drain. As there was high unemployment and the drain had to be constructed through a densely populated area, it was decided that labour-based methods and community contracts would be the most appropriate way to implement the works. The Kalerwe Community Based Drainage Upgrading Project was carried out from April 1993 to March 1994, after which time project management was delegated to the community.

The funding agencies for this project were UNDP, the Government of Uganda, the Kampala City Council (staff and office facilities) and Kalerwe residents (labour). Technical service was provided by the ILO. The funds were channelled through a Project Management Team (PMT) which signed community contracts with Local Project Committees (LPCs). These committees included community members living in a certain area of the Kalerwe settlement. The community contract was assigned to that LPC representing the area the drain was crossing at the time. The community contracts were labour only contracts, and the procurement of material and equipment was carried out by the PMT. The primary drainage channel (2.4 km) was treated as major works. Therefore no community contribution towards construction costs occurred. The secondary channels (1.4 km), considered minor works, involved a direct community contribution of 33% of the total labour costs.

The notable issues in this contract were:

- Although the community was involved in the planning and design, it did not have full control. The capacity building of the community in initiating and planning these types of works was a minor aspect of the project.
- The community as contractor (local project committees) was only responsible for the labour input. There was a clear task division between the Project Management Team (contracting authority) and the contractor (the local project committees).
- The works were carried out effectively and efficiently. However, maintenance has been a problem. During project implementation it was envisaged that maintenance on the main storm water drain would be carried out with decentralized government funds and implemented under community contracts. However, the decentralized government funds were not available directly after the completion of the infrastructure and maintenance has only taken place in an ad hoc manner.

Sources: Tournée & van Esch (2001) *Community Contracts in Urban Infrastructure Works: Practical Lessons from Experience*. Advisory Support, Information Services and Training for Employment-Intensive Infrastructure Development (ASIST-Africa), International Labour Office, Geneva, pp. 23–25.

The third approach is based on local economic development (LED), which is 'a process of participatory planning through partnerships between local government, the business community and the civil society' (ILO, 2004a: 26). LED intends to recognize and realize strategies for territorial development based on the comparative advantages of specific areas. In sum 'the overall objectives of LED are to promote economic development, employment creation and poverty reduction at the local level.' This facilitates negotiation and consensus between the different actors. One example of the role played by local authorities in urban, economic and social development is the FUNACOM program, presented in Box 3.2.

Box 3.2   The FUNACOM program in São Paulo, Brazil

FUNACOM was one aspect of a radical political program introduced by the newly elected *Partido de Trabalhadores* (Workers' Party) in the Municipality of São Paulo between 1989 and 1992. The Workers' Party adopted a comprehensive and participatory approach to the problems of the city, including those of the informal (and illegal) settlements. It encouraged a process of community negotiation in order to develop an action plan for the city, not only with a view to improving the overall environment, but also to achieve a more equitable one. The new action plan comprised an urban policy based on the mobilization and support of local communities. It introduced a new regulatory framework for land use, zoning and building standards. It sought to mobilize financial resources through the restructuring of an existing fund ('the municipal fund to support housing for low-income people' – FUNAPS) and redeployment of municipal resources.

In essence, the program allocated funds directly to the families involved in order to improve housing and infrastructure facilities throughout the city. Families organized themselves into community associations (autonomously functioning legal associations) and were assisted both in the formulation and implementation of local projects by technical assistance teams. The projects, developed in consultation between the community associations and the technical assistance teams, were submitted to FUNACOM for approval. Through this program, the community not only decided on the nature and standards of local projects, but was also responsible for the management and allocation of finances, and participated in the construction process.

Some 'strengths' of the FUNACOM program included:

• The high level of community participation and the decentralized nature of the program ensured that it effectively countered problems of political patronage.

- The localized nature of the project and flexibility of the program resulted in appropriate choices of standards and technologies, according to the preferences, needs and monetary resources of each community.
- The local authority adopted decentralized procedures and accepted local decision-making. The local authority also began to develop public-private partnerships for land development.

The FUNACOM program, as one of the shelter programs initiated by the municipal authority of São Paulo between 1989 and 1992, achieved much greater success in targeting the needs of the urban poor than previous administrations. The program was able to both mobilize local communities to participate effectively and to generate substantial local resources to improve the living and environmental conditions of the urban poor.

Sources: Klink (2006) *The Role of Local Authorities in Promoting Decent Work: Towards an Applied Research Agenda for the Construction and the Urban Development Sector.* Working Paper No. 243, SECTOR, ILO, Geneva, pp. 16–17.
UNCHS (1996) *The Human Settlements Conditions of the World's Urban Poor. UNCHS, Nairobi,* pp. 243–250.
Denaldi (1997) Viable self-management: the FUNACOM housing program of the São Paulo municipality. *Habitat International,* **21** (2), 213–227.

The fourth approach involves the strengthening of urban social dialogue. According to the ILO, conventional processes of social dialogue and collective bargaining usually occur at the national, sectoral or enterprise level between representatives of government and employers' and workers' organizations. Thus 'in the urban context, local authorities can develop their own complementary approaches for an expanded social dialogue, drawing in all the partners at local level' (ILO, 2004a: 25). The aim is to develop a platform which enables the different actors concerned – local authorities, community organizations, formal and informal employers' organizations, local trade unions and workers in the informal sector, along with the Ministry of Labour and other ministries – to share information, consult, negotiate and allocate decision-making.

The fifth approach is the elaboration and implementation of public works and Employment Intensive Infrastructure Programs (EIIP) of the ILO as an instrument of employment generation. EIIP was created in the mid-1970s as part of the ILO response to the deteriorating employment situation in developing countries. A specific example of EIIP in related services is included in Box 3.3.

The principal objective of this approach is to influence investment policies so that they have a greater impact on employment creation and poverty alleviation:

> ## Box 3.3   Selective solid waste collection and recycling in Recife, Brazil
>
> Recife is the capital city of a state in north-eastern Brazil with a population of 1,300,000 people. The city is confronted with poor infrastructure resulting in the limited collection and treatment of domestic sewerage and solid wastes. The contamination of water by wastes and the high incidence of water-related diseases lead to high costs to the public. The municipal institutions therefore turn to social structures and community approaches as alternatives to public services.
>
> The Program of Selective Collection and Recycling of Solid Waste was initiated in 1993. It aims at behavioural change for reducing the production of solid wastes, encourages and promotes the commercialization of recyclable material and stimulates the generation of income. Men, women, children and adolescents work in cooperatives or other community based organizations, generating income through sustainable activities.
>
> The continuity of the initiatives undertaken by the municipality, together with the residents for the Selective Collection of Urban Garbage is ensured by a strong emphasis on environmental education and the high participation of both community-based organizations and the private sector. Through environmental and hygiene education, people learn to separate recyclable materials at the source and donate them to groups which collect, sort and sell them for a living. The difficulties that occurred in the implementation period were caused mainly by the lack of management capacity building and by strong fluctuations in the market price of recyclable materials. Continued partnership between the public and private sectors is an important element of the program. Recyclable materials will continue to be part of the urban waste and industries will continue to be interested in obtaining them.
>
> The results achieved are as follows:
>
> - 73% increase in recycled materials in two years
> - 62% annual increase in volume of material for recycling
> - 482 ton/month reduction in solid waste
> - 56.5% reduction of special operations for waste collection and 2856 dump sites reduced to 124 (43.5% reduction)
> - 5796 tons/month less garbage collected
> - 5 to 20 years expansion of the life of dump sites
>
> *Source*: 'United Nations Earth Summit + 5 Success Stories'
> http://www.un.org/esa/earthsummit/recycle.htm (consulted 21 March 2007)

In most developing countries a high percentage of government investment budgets, as well as gross fixed capital formation, is allocated to infrastructure creation and maintenance. By demonstrating how such infrastructure can be created and maintained in a cost-effective manner with labour-intensive methods, the program has a major impact on creating sustainable employment with available existing resources. Furthermore, by directing such investments towards the needs of low-income groups, the program has a double impact on poverty alleviation, both through the infrastructure itself, and through the employment created during construction and maintenance.

(ILO 2000d: 9)

Therefore, and as highlighted by Klink, employment-intensive infrastructure programs can be considered as 'launch pads' for small contractors to enter into the public construction market. One example of the implementation of EIIP in the provision of community infrastructure is the Maharashtra Employment Guarantee Scheme (EGS) described in Box 3.4. Unfortunately, there is still no evaluation of how local government can generate these processes. Also, it is important to remember that it should not be assumed that this approach automatically generates decent work.

Box 3.4   The Maharashtra Employment Guarantee Scheme (EGS), India

The Employment Guarantee Scheme (EGS) introduced in Maharashtra in the early 1970s was an innovative anti-poverty intervention. The EGS provides a guarantee of employment to all adults above 18 years of age who are willing to do unskilled manual work on a piece rate basis. The scheme is self-targeting in nature. It is totally financed by the State Government. The main objectives of the EGS are to improve household welfare in the short term (through provision of employment) and to contribute to the development of the rural economy in the long term through strengthening rural infrastructure. Works undertaken by the EGS have to be productive. There is an elaborate organizational set-up for the EGS. Initial planning is generally done by the Revenue Department, while implementation is carried out by the Technical Department.

The unique feature of the EGS is its insurance nature. In principle, every person willing to work for the statutory minimum wage can get a job under the EGS. Being permanent, EGS is the major anti-poverty measure in Maharashtra State and relies on 10% to 14% of the State's development budget.

From a modest beginning of only Rs18.8 million in 1973, the scheme expanded to an expenditure of Rs2350 million in 1991 and from 4.5 million person days of employment in the first year to approximately

(Continued)

190 million person days in 1986. This subsequently declined to 80–90 million person days after 1989.

During its lifetime the EGS has, with certain fluctuations, produced between one to two billion workdays of paid employment each year: equivalent to an average of about half a million men and women being at work throughout the year (this means that the EGS has reduced under-employment and unemployment in the state by around 20%).

Some studies have shown that the average period worked per person varies between 25 to 160 days and that the money earned constitutes between one-third and two-thirds of the total income of participating households.

In spite of a variety of difficulties, observers state that none of the other Indian anti-poverty programs have sustained large-scale operation for such a long period or dealt as effectively with corruption and other administrative problems.

Based on estimates at the combined level, it can be concluded that the contribution of the EGS to the total employment/under-employment in the state varies from around less than 10% to 33%. However, the equivalent of 10% to 30% of full-time employment has an impact on a much larger part of that population group because EGS employment is considered only as supplementary or part-time employment.

Sources: ILO (2000d) *Employment-intensive Investment in Infrastructure: Jobs to Build Society.* International Labour Organization, Geneva, p. 49.

Dev, S.M. (1995) India's (Maharashtra) Employment Guarantee Scheme: lessons from long experience. In: J. von Braun (ed.) *Employment for Poverty Reduction and Food Security.* IFPRI, Washington, pp. 108–143.

The sixth approach uses the strategic role of alternative procurement and tendering procedures. In order to promote decent work, local authorities can include additional criteria in traditional tendering contracts, such as the employment of disadvantaged groups and excluded populations. Watermeyer (2006) proposes to evaluate the role of targeted procurement as an instrument of poverty alleviation and job creation in infrastructure projects: 'Targeted procurement provides employment and business opportunities for marginalized/disadvantaged individuals and communities. It enables social objectives to be linked to procurement in a fair, transparent, equitable, competitive and cost effective manner.' Targeted procurement has been used within the South African context, where it has been incorporated in the Constitution of 1996 'to address good governance concerns and to use procurement as an instrument of social policy' (Watermeyer, 2006: 27) and in order to target those groups excluded under the apartheid system. Box 3.5 shows an interesting example from Esmeraldas in Ecuador, in which the local government used public works to generate employment for young people who have been involved in violence and/or crime.

Box 3.5   Esmeraldas: employment for young adults

The high levels of unemployment and under-employment in Esmeraldas have been a major concern and challenge for the local authority. In 2000–2001 the unemployment rate in the city was estimated to be between 68% and 73% of the total working population. These statistics mean that about 70% of the workforce living in Esmeraldas is self-employed and/or active in the informal sector.

Given these conditions, the municipality has encouraged the establishment of new productive activities in the city. It has also ensured that the execution of public sector programs and projects include a component of local labour.

As a preliminary step to authorize the construction of the first large-scale commercial complex by a private business group in the centre of Esmeraldas, the mayor proposed to hire young unskilled labourers who were among the members of local youth gangs. The mayor personally guaranteed that their behaviour and performance would be exemplary.

The results of this initiative were entirely satisfactory both from the point of view of the entrepreneurs and for the young people involved. Many young labourers had the chance to demonstrate their abilities and to learn a trade for the first time in their lives, leaving behind their dangerous criminal life. Nevertheless, and even if the experience was satisfactory, employment was limited in time and did not imply any period of training for workers involved or any commitment from the employers.

Sources: Instituto Nacional de Estadísticas y Censos (2001), Encuesta de Empleo, Desempleo y Subempleo, Área Urbana, INEC, Quito.
Documentation supplied by the Provincial Government of Esmeraldas, and by the Ministry of Labour (Representation in Esmeraldas).

The six approaches identified by Klink (2006) unveil different types of potential for local authorities' interventions in labour markets, particularly in the construction sector. Apart from generic references in the literature mentioned in the first section of this chapter, there is a lack of more detailed studies about the potential of decent work within the construction-related agendas of local authorities. Klink notes that there are ample references about the new productive role of local government within the globalizing economy (resulting in innovative evaluations of local economic development and employment strategies). However, these evaluations are deficient from a decent work perspective. Second, a set of more empirical analyses that focus on the role of local authorities as strategic actors within poverty alleviation strategies does exist. In the more decentralized framework found in Latin American countries, the potential role of local authorities in setting up and applying inclusive policies through safety nets, minimum income programs linked to school attendance and other compensation strategies targeted at the most vulnerable population groups has been highlighted. The noteworthy example of gender equality promotion based on initiatives by the national government of Norway is an innovative example, presented in Box 3.6, which has rarely been used in

---

Box 3.6   Working towards gender equality in employment, Norway

Approximately 43% of employed women in Norway work part time, while men work full time or even overtime. This situation means that there are inequalities in the labour market, as well as in the delegation of responsibility, conditions of income and promotion in career paths. To improve the workforce potential of female workers, men need to share childcare tasks. In workplaces, it should be highlighted that men are not just workers but also parents.

Norway has formulated and implemented public engagements for families with small children. These arrangements make it possible for both parents to combine work with childcare. Flexibility has been reinforced with longer periods of paid leave; leave for fathers has been extended to six weeks. Action has also been taken to facilitate the combination of part-time work with paid parental leave. In addition, the national Government has attributed high priority to providing full-day care coverage, including post-school activities.

In 2007, the Norwegian Government presented at parliament a White Paper entitled *Men and Gender Equality*, the first document of its kind in the world. Summarising this document, Slungärg stated that 'equal parenthood is the key to equality. Equality at work, where both men and women take part in the labour force and contribute their expertise, experience and skills is, in turn, the key to increased productivity and economic growth.'

In order to promote gender equality in the context of working life, the Norwegian Parliament adopted a new law in December 2003 requiring companies to nominate women to at least 40% of board positions.

Sources: Slungärg (2006) The experience of Norway in working towards gender equality in employment. In: *ECOSOC, Full and Productive Employment and Decent Work: Dialogues at the Economic and Social Council*. Department of Economic and Social Affairs. Office for ECOSOC Support and Coordination, United Nations, New York, pp. 130–131.

other so-called developed countries. Broadly speaking, Klink has noted that the majority of innovative examples are related to economic inclusion and community solidarity, without many links to the construction-related services provided by local authorities. The authors of this book have found little evidence to question this conclusion.

3.5   Conclusion

This chapter has separately analyzed three core subjects of this book: decent work in urban areas, the role of local authorities and the context of the construction sector. These core subjects were then explicitly interrelated. Given that there are very few publications on the implementation of decent work by local authorities, this book has tried to bridge this gap and will present specific case studies in the following chapters. Only a few examples of good practice have been found, frequently without reference to the concept of decent work. Still, it has been worthwhile analyzing these concrete case studies. They have illustrated the added value of good practices and helped understand why problems are found in some projects and programs initiated by local authorities. From this perspective, recommendations for future practice are included in the last chapter of this book.

Note

Chapter opening image: A woman carrying more than 30kg of stone slabs on her head for a construction site in Indonesia. Photograph courtesy of J. Maillard, ILO.

(1)   www.ilo.org/ipec

4 Bulawayo

Beacon Mbiba and Michael Ndubiwa

4.1 Introduction

This chapter presents the case study of Bulawayo, Zimbabwe. It begins with an overview of the political context and an examination of economic development at both the national and local levels. Qualitative information, mainly derived from interviews with key actors from different institutions involved with the construction sector, has been analyzed. The fieldwork also examined efforts made by the local authority to promote employment in the construction sector and related services, and to assess how the city of Bulawayo has been able to apply the principles of decent work through examples of best practice.

4.2   National, regional and local context

4.2.1   The national context

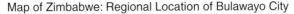

Map of Zimbabwe: Regional Location of Bulawayo City

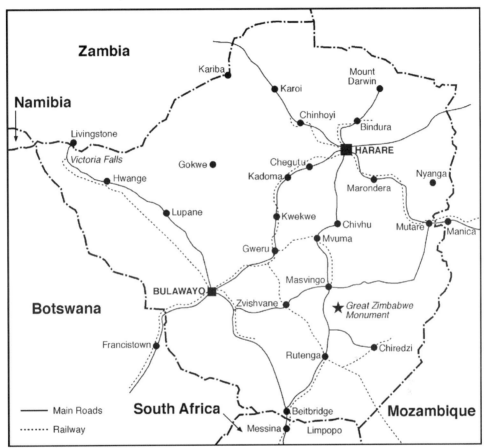

Change and continuity: beyond colonial heritage

Although at independence in 1980, Zimbabwe adopted a socialist policy outlook, it remained a fundamentally capitalist economy (GoZ, 1982). It has achieved world acclaim in smallholder food production (Eicher, 2003; Gabre-Madhin & Haggblade, 2004; Rukuni & Eicher, 1994), education and health (Raftopolous, 2004), infrastructure and decent work provision for the previously marginalized majority of black Zimbabweans (LEDRIZ, 2005). Zimbabwe was also applauded globally for its 'national reconciliation policy' adopted in 1980, its role in brokering peace in Mozambique, its SADC leadership and its leading role to end apartheid in Namibia and South Africa. Zimbabwe has been influential globally, for example in UN peacekeeping missions in the 1980s and early 1990s in Somalia, Liberia and the Balkans. However, contradictions characterized the economy; a factor compounded by continued inequitable land distribution, obsolete industrial equipment and production methods that have not been able compete in emerging global markets (Bond & Manyanya, 2002).

In an attempt to reverse the economic decline that set in during the mid-1980s, a reform program, The Economic Structural Adjustment Program (ESAP), was adopted in 1991 (Bond & Manyanya, 2002) with a view to opening up the economy, reducing government expenditure and removing tariffs, etc. However, while the service and tourism sectors experienced positive growth, increased productivity did not materialize, notably in the manufacturing sector (Zaaijer, 1998). Instead a rapid increase in income and consumption inequalities set in, leading to political ferment by the mid-1990s. At this point, President Robert Mugabe and the ruling ZANU (PF) party abandoned any remaining aspects of ESAP market economics promoted by IMF and the World Bank, in favour of pragmatic dictates. In particular, a combative approach associated with revolutionary language became the norm and a land distribution programme, the 'Third Chimurenga', was fast-tracked to the policy forefront from 1995 onwards on the back of the 1993/94 Rukuni Land Tenure Commission report.

Attempts at dialogue with donors, labour movements, employers and the intelligentsia, initiated between 1997 and 1998 in an attempt to salvage the situation, collapsed. By the time of the February 2000 referendum for a new national constitution (which was rejected by 54.7% of the electorate), the political atmosphere was completely changed and had become poisoned. The Movement for Democratic Change (MDC), a trade union and commercial farmer-anchored political party, de-campaigned the draft constitution and came close to defeating the ruling ZANU (PF) at the 2000 Parliamentary elections, in which the MDC got 57 seats against 62 for the ruling party. Charges of alleged vote rigging, violence, voter intimidation and human rights abuses in general were made by the MDC, international human rights groups, the EU, Australia and the USA. On this basis, since 2001 the EU, Australia and the USA have consolidated the isolation of Zimbabwe (Patel & Chan, 2006), imposing targeted sanctions, blocking loans to the country (e.g. from IMF, Inter-American Development Bank, African Development Bank, Inter-American Corporation and the World Bank) and freezing all bilateral aid, except in extreme humanitarian situations. It has become increasingly difficult for Zimbabwean firms to do business in the USA and EU, and vice versa.

USA economic sanctions on Zimbabwe are spelt out *inter alia* in section 4(c) of the Zimbabwe Democracy and Economic Recovery Act (2001). Lack of transparency in how these countries compile their 'smart sanctions' lists was highlighted by Australia's 2005 list which included people who were dead, incorrect details and individuals not associated with the Government; following protests, 53 were subsequently removed from the list of 127.

Of late, escalating inflation and a sharp decline in GDP have emerged as key economic challenges. A rapid rise in inflation since 2000 and misaligned exchange rates make it difficult to calculate accurate US$ GDP figures from nominal GDP values in Zimbabwe dollars (EIU, June 2005: 27). Therefore, while GDP has risen sharply in nominal terms, the reality is the opposite in real terms (CB Richard Ellis, 2006; Intermarket Research, 2004). After political tension, commodity shortages, foreign currency and food, fuel and electricity shortages, an inflation rate of over 1000% by end of April 2006 and three million percent by mid-2008 this has been acknowledged as one of the biggest economic challenges the country now faces.

The Third Chimurenga and the 'Look East' foreign and economic policy

Patel & Chan (2006: 176–177) argue that there is an organic link between Zimbabwe's armed struggle for independence (known as the Second Chimurenga) and the way Zimbabwe has fiercely guarded its sovereignty, even at great cost in recent times. Having survived military and economic destabilization by apartheid South Africa as well as sanctions threats from the UK and USA in the 1980s, Zimbabwe now finds itself facing real economic threats as a result of the latter. Since 1997, it has recast defence of its sovereignty against these threats in the language of the Third Chimurenga, in which a domestic fast-track land reform is a key component which has brought about economic destabilization in the short term.

The fast-track land reform program basically converted what were formerly white-owned commercial farms which earned foreign currency into 'nationalized', production enterprises as part of the peasant system[1]. Agriculture, the backbone of the economy, was decimated. Combined with political upheavals and economic isolation, the Third Chimurenga has been followed by rapid economic decline, further political instability, impoverishment, unprecedented emigration and the collapse of social services, health, education and transport. The human development gains from the first decade of independence have been reversed (LEDRIZ, 2005). Economic decline has had a negative impact on any attempts to revive the agricultural sector. Combined with the impact of HIV/AIDS on the economy, both industrial and agricultural productivity have continued to fall. Other problems surface continuously; shortages of fuel, food (e.g. bread, sugar and cooking oil), electricity, transport, and foreign and even local currency.

The response of ZANU (PF) has seen a return to the 'old political guard' (and mistrust of the 'young Turks' within the party); a militarized and uncompromising 'sledge hammer' approach to domestic politics and economic management. This is a vitriolic anti-imperialist stance at the global level and a 'go-it-alone' attitude forming the backdrop for the 'look east policy' aimed at the 'dispersal of dependence' (Patel & Chan, 2006: 176). The 'look east policy' is characterized by opening up and extending economic, military and diplomatic relations with countries in the East, particularly Malaysia and China, who supported Zimbabwe's struggle for independence prior to 1980. China–Zimbabwe relations have helped to keep key sectors afloat: Air Zimbabwe, the airforce of Zimbabwe, ZUPCO (the national bus company) along with mining, building materials production and the construction of national projects, such as the Harare Norton Motorway[2]. But this has not been able to reverse economic decline in the short term. For the economy to stabilize and recover, the Chinese investments may need to be complemented by a comprehensive return of western resources in the mining and industrial sectors. Capital flight, characterized by ordinary Zimbabweans taking their savings elsewhere, as well as a negative image of the country held by many in the western world, needs to be reversed.

Employment and the construction sector

According to the Economist Intelligence Unit (2003: 23), the proportion of the population formally employed in the economy as a whole dropped from 18% in 1965 to under 10% in 2000. It is therefore understood that the rest of the population is employed in the informal and peasant sectors (LEDRIZ, 2005). Second, employment since 1980 has made a marked shift away from manufacturing to sectors such as education, health and financial services. Agriculture is reported as employing 26% of the formal labour force, as compared to 15% in the manufacturing sector (EIU, 2003: 23).

In terms of share of GDP, the construction sector witnessed a boom in the 1970s and a general decline in the 1980s and 1990s, with share of GDP falling from 5%, in the 1970s to 2.5% in the 1990s and 2% in 2001 (EIU, 2003: 30). However, there have also been some short growth periods, such as the one between 1996 and 1998. The relatively small contribution of construction to the formal economy can also be seen in terms of employment. Just over 80,000, or 5%, of the total formal employment in 1998 was in the construction sector. This fell from about 100,000 workers in 1990 to 90,500 in 1993; 71,800 in 1995; 78,100 in 1997; 53,800 in 2000 and 39,300 by 2003 (CSO, 1998: 11; 2003: 11). Major construction employers include contractors such as International Holdings, Costain, John Sisk and Sons, and Gulliver Consolidated. There are many informal contractors, several operating as sole agents.

The construction sector has been hard hit by economic problems as these have led to high constructions costs (see Table 4.1). The rapid rise in costs can be observed in the cost index (last column in Table 4.1) or the detailed price indices for civil engineering and building materials (see Table 4.2). Prices escalations were a futile attempt for firms to reverse serious

Table 4.1   Construction (building) cost increases, 1995–2004 (Z$/m^2)

Year	Standard house	Standard factory	Standard office block	Arithmetic mean	Index: 1995 = 100
1995	2300	1300	2500	2033	100
1996	2650	1675	3000	2442	120
1997	3530	2270	4720	3507	172
1998	4950	3190	6620	4920	242
1999	8915	5760	12,040	8905	438
2000	17,025	10,978	17,275	15,093	742
Mid-2001	29,200	18,000	30,950	26,050	1281
Mid-2002	56,000	34,000	59,350	49,783	2448
Mid-2003	800,000	550,000	750,000	700,000	34,426
Feb 2004	1,750,000	1,200,000	1,650,000	1,533,333	75,410

Source: Robertson Economic Information Services, Harare, Zimbabwe, 2006.

liquidity problems. However, the cycle of rises and reduced demand led to reduced production closures.

Cement production, a key component in the construction sector, has been paralyzed by closures or reduced capacity of the main cement plants[3]. Problems arise due to fuel shortages or lack of spare parts, affecting coal production and the capacity of the National Railways of Zimbabwe (NRZ) to transport coal to cement plants. Thus, in 2003, one after the other, the Sino-Zimbabwe Plant, Portland Cement, Unicem and Circle Cement reduced or closed operations for long periods[4]. The cement shortages soon after 2000 hit the construction sector, which had been booming a few years before. Some media reports projected a recovery after 2005[5].

The backward and forward links in the economy create a domino effect, experienced not only in the construction sector but also throughout other sectors of the economy. The fertilizer production industry, for example, has been equally affected, with fertilizer shortages affecting agricultural productivity, even in periods of plentiful rain, such as 2005–2006. With construction costs rising, both institutional and individual clients have found it more and more difficult to construct new buildings or infrastructure. This is the environment within which local authorities need to initiate local economic development and promote decent work.

Local government in Zimbabwe

Local authorities in Zimbabwe primarily implement powers given to them by the Ministry of Local Government and Housing. For the administration of urban areas, this is done in line with The Urban Councils Act [214] and the Regional Town and Country Planning Act [Chap. 29: 12] and all its allied statutes. The Urban Councils Act [214] provides for the establishment of urban local authorities, whose categories range from the smallest area

Table 4.2   Civil engineering and building materials price index (1990 = 100)

	Water stops	Cement	River sand	Crushed stone	Bricks	Diesel	Structural steel	Plant	Reinforcing steel	Concrete pipes	Bitumen
1990	100	100	100	100	100	100	100	100	100	100	100
1991	106.4	218.4	177.0	147.2	128.0	135.8	160.9	158.7	182.5	149.2	100
1992	147.0	303.2	177.8	184.9	157.7	205.0	303.1	227.5	300.0	207.7	100
1993	190.6	366.5	183.2	262.6	178.6	250.8	350.9	266.5	366.4	288.3	100
1994	191.2	420.8	194.7	297.3	197.9	321.8	428.1	374.4	458.0	288.3	100
1995	246.8	427.9	230.4	329.0	246.1	381.8	519.1	417.4	550.0	302.7	100
1996	249.6	590.4	255.8	404.0	341.4	471.7	632.0	486.2	647.4	325.5	101.1
1997	249.6	903.2	295.2	550.5	457.1	609.6	803.7	565.3	772.4	326.7	137.5
1998	249.6	1123.8	337.8	691.4	513.8	875.2	1220.6	1199.5	1143.7	386.4	154.4
1999	245.6	2005.2	626.8	1117.7	837.2	1641.9	1966.3	1894.1	2001.3	610.7	237.2
2000	374.0	2760.7	1461.9	1986.0	1572.3	4065.4	2574.8	2207.2	2396.9	923.4	253.1
2001	801.2	4648.3	2226.9	3559.1	2406.6	8479.4	4376.4	5763.7	4130.3	2152.2	719.9
2002	3718.3	8777.6	3802.4	10185.5	5453.2	10516.4	11806.1	12290.6	17503.3	4792.3	2269.9
2003	21072.5	80641.0	55722.9	85707.8	47092.6	59496.6	116509.4	107120.0	153451.6	22920.2	8330.6
2004	76920.4	536741.4	253414.5	448508.2	224081.4	506020.8	470732.0	347373.5	490032.2	147093.0	26625.4
2005 Jan	95910.3	723520.4	875812.2	1495370.7	356785.7	590053.5	759950.1	429149.5	1014495.4	328415.6	40302.7
2006 Jan	2601941.3	9736518.0	13628709.7	20682921.8	5301340.3	3294795.2	8864052.2	7490543.4	15361100.3	7308986.9	1048353.3

Source: Central Statistical Office (CSO) *Civil Engineering Price Index*, Harare, 10 March 2006.

boards, through the local boards, town councils and municipal councils to the largest category of city councils. Local authorities can raise revenues and provide services in areas of their jurisdiction.

As with the national economy, the Government's priority soon after independence was democratization and de-racialization: to create unified or unitary local authorities in both rural and urban areas with public sector investment largely targeted at the formerly neglected rural communal lands – home to 70% of the national population. Rural local authorities have fewer resources and less autonomy than urban local authorities. In both cases, local authorities are run by an executive consisting of bureaucrats (responsible for health, planning, housing, finance, works and transport, etc.) on the one hand and a political council made up of members elected by residents on the other. Large cities and metropolitan areas also have an executive mayor elected by residents from the area[6].

Mutizwa-Mangiza (1991) concluded that, unlike in larger parts of Africa, the administration of urban local authorities in Zimbabwe in general was sound and that of Bulawayo in particular was encouraging and exemplary. Urban local authorities have a relatively large degree of autonomy in terms of finances and organization. Bulawayo, in particular, has earned a reputation nationally and internationally as a well managed city (Hamilton & Ndubiwa, 1994; Zaaijer, 1998).

While urban local authorities can raise revenue and provide services in their areas of jurisdiction, there are other institutional entities that operate as local authorities in the same areas, such as parastatals like the Zimbabwe Republic Police (ZRP), the Zimbabwe National Water Authority (Zinwa)[7] and the Zimbabwe Electricity Supply Authority (ZESA), which in 1986 took over the supply and administration of electricity from city councils. Elsewhere, institutions like mining companies and other parastatals may operate as local authorities in some instances. Social security issues within a local authority area, like Bulawayo, for example, are administered by NSSA and in this case Bulawayo has little input into its management.

Key institutional changes since the mid-1990s have seen central government putting more controls on how the urban local authorities can implement the Urban Councils Act. In particular, the introduction of the executive mayor in the late 1990s had meant that decisions that would have previously been made by consensus of the council based on advice from the technical heads of departments could now be overridden by the executive mayor in liaison with the minister. The resulting tensions led to the phenomenon of 'commissions', appointed by the minister of local government to run the affairs of council in place of the elected councillors and mayor. The instability and uncertainty in Harare's governance is illustrated by continuous conflicts at the top; heads of departments are continuously hired and fired or hold posts in an acting capacity. For instance, top management in 2006 comprised of: acting chamber secretary, acting human resources manager, acting director of housing and community services, acting director of waste management, acting director of finance, acting director of works and acting director of health services. While this has been the case in Mutare, Chitungwiza, Norton and especially in Harare, the capital, Bulawayo has been able to 'survive' this unpleasant scourge (see Sachikonye, 2006; Zaaijer, 1998).

The ruling party ZANU (PF) has consistently lost the urban vote in national and urban local elections since 2000, leading to a highly contested political environment. In a bid to exert greater management control over the two metropolitan areas of Harare and Bulawayo, the Government introduced the office of a resident minister in charge of the area. What this means is that the mayor is no longer the highest local government politician in the area, but the resident minister. In addition, while reporting to the Minister of Local Government and Housing, the council and the mayor must also report to the resident minister.

Thus, above the category of city council, we now have a 'metropolitan authority' appointed by the central government, not elected by the people. This basically means that the degree of autonomy that existed in the 1980s (Mutizwa-Mangiza, 1991) has now been significantly eroded. Through the resident minister and the commission, appointed by the minister to run the affairs of Harare and Chitungwiza[8] since the late 1990s, there is generally more centralization than decentralization of power in urban administration. For Bulawayo, potential tensions and conflicts arising from this arrangement have been avoided, partly because of the city's distance from national political power struggles as well as the culture of the administration, which prioritizes service to the people over politics (interview with Mayor of Bulawayo, April 2006).

4.2.2   Regional and local context

Bulawayo, the city of kings

Bulawayo's modern history can be traced back to the 1890s when it was transformed from the Ndebele settlement of grass thatched huts to the regional city it is today (Hamilton & Ndubiwa, 1994). It is located in a dry agro-ecological region whose hinterland covers Zimbabwe's Midlands, Matabeleland South and Matabeleland North provinces, as well as large parts of Botswana, South Africa and Zambia. Wildlife and mineral wealth are key natural resources in this hinterland. There have been major finds in methane gas in the Lupane and Hwange areas (Bulawayo City, 2000: 7). It was the industrial capital of Zimbabwe in the early years before Harare took over in the 1970s. It remains the headquarters of the Railways of Zimbabwe (the second largest employer after the Government) and boasts railway links to Botswana, South Africa, Mozambique and Zambia, leading to the Democratic Republic of Congo and Angola. The Joshua Muqabuko

Nkomo International Airport provides links to tourist destinations such as the Victoria Falls and Hwange National Park, as well as regional resorts in Namibia, Botswana and South Africa.

At its industrial peak and before the advent of the Economic Structural Adjustment Programs of the early 1990s, Bulawayo had a diverse industrial base and comprehensive engineering industry that included textiles[9], radio manufacturing (the largest in Africa) a tyre factory, and hide and meat processing. It now has a range of training facilities and services for railways and the mining industry, and is home to the National Free Library, the large Mpilo Central Hospitals, a stadium able to host international games and the National University of Science and Technology (partly still under construction since opening in 1991). Bulawayo is also home to the Zimbabwe International Trade Fair (laid out over twenty-five hectares), through which the country can showcase its industries to the world. Bulawayo is a cultural centre, not only in terms of Ndebele traditions, but nationally and regionally in terms of the labour and nationalist movements whose formative years were there. Its hinterland (in places like the Matopo Hills, Gwaai catchment, Shangani area and Khami Ruins) bears witness to Zimbabwe's history of violence, struggle and pride.

Administratively, Bulawayo rose from a sanitary board in 1894 and municipal council in 1897 to a city in 1943 (Hamilton & Ndubiwa, 1994). The areas around Bulawayo are administered by five rural local authorities (Bulawayo City, 2000) including Umguza, Umzingwane and the Department of National Parks and Wildlife Management, which administers the wildlife areas. The provision of water to the city has been a constant problem from as early as the 1980s, and the drought of 1991/1992 was a real threat to the sustainability of the city. The Government and the city have an ongoing programme to construct the Gwaai-Shangani Dam, some 350 km to the north of the city, from which water will be conveyed by pipelines. A long-term solution to the city's water problems in the form of a pipeline from the Zambezi River has thus far remained a strictly theoretical proposition, due not only to the huge cost and alleged local politics involved, but also the regional geo-political and environmental concerns that need to be resolved. Associated with these water problems is the issue of sewers that are now overloaded; although new works are planned or under way (Bulawayo City, 2000: 9).

Further threats to Bulawayo's sustainability are the national economic conditions described elsewhere in this report, and the HIV/AIDS epidemic (Bulawayo City, 2000), which has put a strain on household and enterprise economies. Thus, the revenue base of Bulawayo is under threat. In addition, the harsh economic climate has seen default in payments of rates, not only by households and firms but also by central government departments – with the council owed up to Z$450 million in 2000 (Bulawayo City, 2000: 12)[10]. The bulk of the council's recurrent expenditure (about 61%) is spent on salaries.

According to the CSO census figures, while the city's population is growing, the rate of growth has slowed from 5.9% in the 1970s, to 4.5% between 1982 and 1992, to between 3.1% and 2.4% during the 1992–1997 period and around 2% in the post-2000 period. This is largely due to reduced economic opportunities and lower rural-to-urban migration. Instead, those not migrating to seek income opportunities in Harare have looked to Botswana and South Africa, especially after 1982. HIV/AIDS, as well as general decline in fertility rates, has also contributed to this decline in growth rates. Consequently, Bulawayo's projected population figures of 1,011,037 in 2000; 1,184,637 in 2005 and 1,562,905 in 2015 have turned out to be higher than the census returns, for example the census results of 2002 gave a total population of just 700,000 (CSO, 2002).

The preceding sections have provided a panoramic view of local authorities in Zimbabwe and how Bulawayo operates in a challenging economic climate. The institutional and political challenges facing the city (such as water) have also been signposted. Zaaijer (1998: 1) captured concerns that while Bulawayo's conservatism and excessive preoccupation with 'rules' are the reasons behind its status of a 'best managed' city, such an administrative character may stifle innovations. The sections that follow will demonstrate that, contrary to such fears, Bulawayo has achieved strategic innovations in the urban development sector; successes with direct positive impacts on construction activity and employment creation.

Key concerns in Bulawayo's local economic development are droughts and the water crisis, as well as the ESAP policy of the 1990s. ESAP introduced institutional reforms and economic restructuring (Bond & Manyanya, 2002) that led to a decline in industrial output from 32% in 1979 to 25% in 1995 (Zaaijer, 1998: 23). In particular, the opening up of the economy led to inflows of cheaper goods from global markets, forcing local companies either to close or lay off workers (Intermarket Research, 2004: 14–15 and 28). Key informants are always quick to mention the 1000 workers retrenched at Merlin in 1995 and that over 10,000 workers lost their jobs in the 1990s alone. CSO (2004: 31) figures confirm that this indicated a decline in formal economy jobs from 153,200 in 1989 to 134,500 in 2002 – close to 20,000 jobs lost in the city over a twelve-year period. Most of these workers either sought work in South Africa or went back to the rural areas. Those who stayed in Bulawayo found themselves in the informal economy. By the mid-1990s, up to 40% of the working age population was in the informal economy. Activities in this sector followed the national pattern of shifts from manufacturing to services, retailing and trade (Kanyenze *et al.*, 2003).

Local economic development in Bulawayo

This section pursues the question of key actors, policies and programs involved in local economic development, employment creation and responses to decent work challenges. In particular, it considers the role of Bulawayo City in strategic planning and the development process around which significant construction activity takes place. It shows how partnerships between the city and the private sector are central to the success witnessed in the city. As already noted, Bulawayo has been championed nationally and internationally for its unique physical and social outlook (Hamilton & Ndubiwa, 1994; Mutizwa-Mangiza, 1991; Zaijer, 1998). It is a well planned and managed city driven by a deep-seated City Hall historical culture geared to serve the people, irrespective of political or social orientation. While the attractive and harmonious environment is easily noticed, less noticeable are the strategic decisions made by the city's leaders. A number of these qualify as best practices.

In 1982, Bulawayo was Zimbabwe's first post-independence city to produce a city master plan in accordance with the Regional Town and Country Planning Act (1976). This served as a framework for the city's infrastructure development, housing delivery and service provision. It enabled the city to accommodate the huge influx of rural migrants following the abolition in 1980 of prohibitive colonial structures, which had restricted the movement and urban citizenship of the African population. In another demonstration of its commitment to planned development, Bulawayo was the first to initiate revisions to its master plan, culminating in the 2000 to 2015 Master Plan, adopted in 2000 in accordance with section 20 of the Regional Town and Country Planning Act (1996 revised edition). Unique in the revision process and its outcomes, was a conscious integration of spatial strategic thinking on the one hand, and organizational strategy, mobilization of resources, budgeting, strategic socio-economic planning and stakeholder involvement on the other. Urban agriculture is now an integral part of local economic development planning for the city.

Bulawayo's 2000–2015 Master Plan took the lead (in Zimbabwe) in transforming city planning from a purely land-use, physical-planning endeavour into a broader, more responsive and dynamic strategic activity that turns the plan into a corporate management tool. As with the 1982 version, the new plan aims to ensure sufficient land for commercial, industrial, residential and institutional development, to enhance the investment climate and stimulate local economic development, and to act as a framework for employment creation and social, recreational and cultural activities. Bulawayo's management is in this way always ten years ahead of events. For example, at any time, the city has planned and serviced land to accommodate thousands of new dwellings. Over the past five years when resources for bulk infrastructure development have been hard to get, the city has adopted infill development strategies.

To summarize, Bulawayo was the first city in Zimbabwe to produce a master plan, the first to revise its master plan and the first to produce a strategic plan and corporate strategy. At any one time in the 1990s, the city had a stock of up to 30,000 planned and serviced stands for

---

> **Box 4.1   Indicators of Bulawayo City's success in delivery of low-income housing**
>
> • Between the 1989/1990 and 1993/1994 financial year, 32,750 low-income dwelling units were constructed to accommodate over 200,000 people at an average occupancy rate of six persons per dwelling (Bulawayo City, 1995).
> • For the five years prior to 2005, a total of 25,000 houses were constructed to house more than 120,000 people (city planner, interview April 2006).

residential development. Currently, despite the tough economic environment, Bulawayo has serviced land for low-income residential development, including 1000 stands in Pumula South, 10,000 in Cowdray Park and 7000 in Emganwini. According to the city planner, surveys that used to be two years ahead of time are now close to three years ahead. This has been achieved through use of joint private sector and local authority survey teams. It is expected that new land for up to 43,000 residential stands will be available by the end of 2006 in the area between the Joshua Mqabuko Nkomo Airport and Trenance. The main barrier to be overcome, which has been a consistent one over the years, is the provision of off-site infrastructure, especially water and sewerage.

The City Master Plan 2000–2015 sets aside land for the establishment of major water treatment works to the north of the city to process and manage water when Government investments in the planned Gwaai-Shangani pipeline materialize (Bulawayo City, 2000: 114–115). See Box 4.1.

Partnerships for local economic development

Bulawayo's success in implementing its strategic plans has been achieved through a combination of delivery models that enable partnerships among stakeholders. In the mid-1990s, Bulawayo signed partnership agreements with construction investors from Malaysia. Promoted by the central government, this saw the creation of Zimbabwe–Malaysia Holdings (ZIMAL Holdings) to produce bricks and construct up to 15,000 housing units in Cowdray Park. The progress on this is given in Appendix 4.1 at the end of this chapter. Bulawayo was thus a direct beneficiary of the Government's 'look east' policy long before the post-2000 economic crisis.

Other partnerships have included employers providing employer-assisted accommodation, such as the Zimbabwe National Army in the past, or CABS Building Society's ongoing project in Pumula South; international development donors, such as USAID and the World Bank (especially in the 1980s and 1990s); Building Brigades, Self-Help; private sector financiers and construction companies (both large and small) and the Government.

A recent example is the housing development partnership that saw private companies, the city and individual households constructing an agreed numbers of units with financial support from the Reserve Bank of Zimbabwe (RBZ). Traditionally the RBZ does not participate in such projects. However, given the economic environment, where local authorities do not have enough resources for development projects, combined with the bank's need to promote inflows of diaspora foreign currency, the RBZ put forward Z$8.3 billion to service 119 stands in Parklands East and 245 stands in Mahatshula. These have been allocated to home seekers on the waiting list for development. The funds also supported servicing of another 101 stands at Emganwinini II Millennium Housing Scheme, where construction is complete.

Bulawayo's strategic planning and management framework has enhanced a consistently developing investment climate, despite national and global economic constraints. The construction of housing and related urban development activities has generated both direct and indirect employment. Although no employment figures are available, it is clear that even where the local authority itself is not involved in construction, the availability of serviced land has enabled many private sector and informal operators to create employment in the city.

Bulawayo City incentives to promote investment and employment creation

As part of its long-term strategy from the early 1990s, Bulawayo put in place a package of incentives to encourage indigenous black Zimbabweans to invest in commerce and industry –areas they had been excluded from for decades – to attract new sustainable industry and commerce and to consolidate the operations of existing investors. Incentives covered access to land, tax rebates and holidays, special tariffs for water and service charges, guaranteed project approval times, waivers of plan-approval fees and speedy allocation of housing land to workers associated with the investing companies.

It is crucial to emphasize that this package of incentives was not a *bambazonke* – siNdebele term meaning he/she who grabs everything. This package of incentives was designed not to be 'for everyone and everything' but was strategically targeted. There was a clear set of criteria to discriminate in favour of preferred investors and developers; criteria based on the number of people to be employed, size of the investment in terms of dollars, new technology to be introduced, participation of local or indigenous persons, export orientation of the industry and pioneers. The way the incentives would work is described in Appendix 4.2 at the end of this chapter. The incentives could be provided in different combinations and the table structure makes it feasible for different sections of the local authority to review their set of inputs into the package. The table is organized as follows: one chooses a developer (first column on the left) then moves across the row to check existing incentives available to the investor. For example, local/indigenous persons (row o) have a lease option available (column 3) as well as professional services and incubation support. In the event that the local/indigenous investor happens to be an employer of say 200 people, then he/she would be able to purchase land from the city at 80% of the original cost (intersection of row b and column 2). Assuming that she/he had an investment worth between 5–10 million dollars, then a two-year tax holiday would apply, as well as guaranteed approval times of 90 days for the project application and 30 days for the plan application, plan fees at 60% of the original price and availability of electricity connections to the development site.

These time incentives are noteworthy given that Zimbabwe has had some of the slowest turn-around times for land-use planning, surveys and building plan approval; it has an overall time of forty months, compared to the sub-Sahara average of about twenty-three months and global average of about twelve (GoZ/USAID, 1994: 28). Ordinarily, processing of the building application alone would take 3.75 months for Bulawayo, compared to 5.75 months for Harare. Clearly, with such delays and in the absence of special package and preferential treatment, the cost of doing business with and investing in cities in Zimbabwe is prohibitive. Thus, it is this package of incentives that has led to sustained housing and urban development in Bulawayo, even in lean years.

The construction sector in Bulawayo

The vibrancy of the construction sector is considered a good indicator of the 'health status' of a nation or city's economy. Using the values of building plans approved over time in Harare and Bulawayo, it is clear that the construction sector has seen a gradual decline since 1989/1990, although there have been periodic booms like that in the mid to late 1990s. While nominal values of plans approved and actual work done have seen dramatic growth, in real terms the decline is significant (compare Figure 4.1 against Figure 4.2).

Computations based on construction data (CSO, 2001) show that the value of work done in 1999/2000 (excluding repairs and alterations) was 6% (Harare) and 11% (Bulawayo) of similar work done in 1989/1990. The cost of materials has risen to levels that make building unprofitable as a business and unaffordable for many households. This is associated with a reduced supply of materials for both local and imported components.

There are also a number of further trends and changes in the status of construction in Zimbabwe that can be exemplified by the Bulawayo scenario:

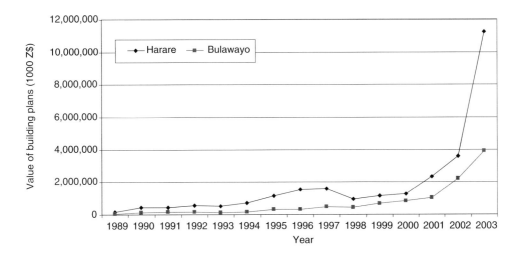

Figure 4.1   Building plans approved in Harare and Bulawayo, values in thousands of Zimbabwe dollars (nominal values)
Sources: Original data from *CSO Construction Bulletin 2001* (p. 19) and *Construction Bulletin 2003* (p. 41), Government of Zimbabwe, Harare.

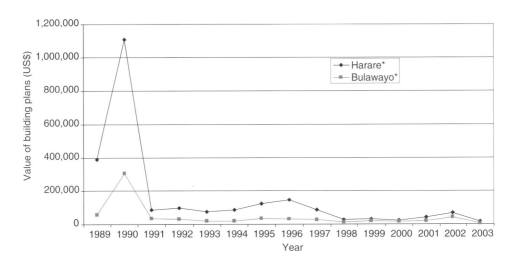

Figure 4.2   Building plans approved in Harare and Bulawayo, values in thousands of US$
Sources: Original data from *CSO Construction Bulletin 2001* (p. 19) and *Construction Bulletin 2003* (p. 41), Government of Zimbabwe, Harare.

Figure 4.3   Repair and alterations April 2006; formerly Downings Bakery building to be Bulawayo HQ for the mobile phone company ECONET Pvt (Ltd). Corner of Robert Mugabe and Leopold Takawira Avenue (facing north east). Photograph courtesy of B Mbiba.

- A switch from new construction projects towards repairs, alterations and maintenance. Construction in Bulawayo CBD dried up at the end of the 1990s and only small repairs and alterations are visible (Figure 4.3). Large public sector projects, like the National University of Science and Technology, are also affected by the unavailability of materials and high costs, leaving buildings unfinished for years.
- The decline in the economy and shrinking of construction have resulted in the exodus of construction firms and qualified bricklayers, artisans, carpenters and others to other countries in the region, notably Mozambique, South Africa and Botswana. According to key informants, this has led to shortages of such skills, allowing the few remaining to demand high fees for their work. Individuals have moved from the formal sector to the informal, while some companies have closed shop, scaled down or diversified into other activities.
- Figure 4.4 shows the decline in residential construction activity since 1997. This is the sector with the capacity to employ many people, particularly those in the informal category. The 1990s boom in low-cost

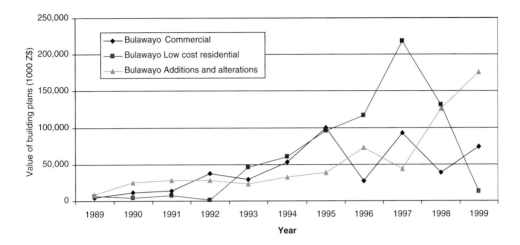

Figure 4.4   Building plans approved in Bulawayo
Source: Data from *CSO Construction Bulletin 2001* (p. 19, nominal values), Government of Zimbabwe, Harare.

housing has subsided due to the reduction in public sector funding, but remains an area of activity in Bulawayo's western suburbs. Combined with repairs and alterations, new development is stimulated by resources from the diaspora, especially those who live and work in South Africa and popularly referred to as *injiva*.

Injiva construction activities are concentrated in the housing sector, largely to accommodate family members, rent out or as an investment for income after retirement. The Reserve Bank of Zimbabwe has initiated a Homelink project designed to help those parts of the diaspora to build or buy houses, although not every *injiva* takes advantage of this possibility. *Injiva* house construction (reportedly two in every five houses under construction) supports a vibrant informal construction economy. Builders and artisans in the informal sector have good skills, most of which were acquired in the formal industry both in and outside the construction sector. A detailed assessment of the nature of the informal construction sector and its broader economic links would be a valuable exercise in the future.

The ongoing Government housing programme Hlalani Kuhle has also boosted the construction sector in Bulawayo. This is a housing programme initiated for those whose shelters were destroyed in 2005 by the Government's Operation Murambaswina. In Bulawayo, as elsewhere, the Zimbabwe National Army is a lead agent in the construction of Hlalani Kuhle houses, although Bulawayo equipment and workers are also involved. Unfortunately, sensitivities surrounding this programme made it impossible to interview representatives of agencies or workers involved in the construction; our team was barred from entering construction sites.

There is also some private sector construction activity, such as the construction of houses in Pumula South, which was funded by CABS Building Society. The largest project with private sector involvement, however, is the Cowdray Park development, which is in turn linked to Bulawayo's long-term strategic plans to construct thousands of middle and low-income houses.

4.3   Decent work indicators

This section presents the results of the statistical compilation of key indicators, in particular those regarding the four components of decent work: employment, social security, workers' rights and social dialogue. The focus is largely on the formal sector, although some respondents have argued that formal sector figures equate to the informal, since from the mid-1990s the whole economy has been 'informalized'. Where possible, the role of key agents in promoting decent work, and notably the contribution of the city of Bulawayo, will be highlighted. However, the main objective of this chapter is to provide comparative indicators. Rather than presenting the information and the analysis separately, the approach is to offer some analysis and possible conclusions immediately after each set of primary data is presented.

4.3.1   Indicators of employment

Unemployment

The unemployment rate is calculated by measuring the proportion of the population aged 15 and above that was unable to find work in a given period. The trends for the country are summarized in Table 4.3. The construction sector unemployment rates are low at both the national and local levels, although the local level shows a significant increase in unemployment conditions from 2000 to 2005. The remainder of this section will explain why these figures are much lower than those reported by the media and non-government organizations (NGOs).

In the 1999–2000 period, 230,463 of those in the working age group were unemployed throughout the country and 40,837 at the Bulawayo City level. The Census Report (2002) shows regional variations in unemployment rates, with Bulawayo recording the highest level at 25.29%, followed by Harare at 18.12%. These regional variations are described in Figure 4.5 and show that the unemployment rate in Bulawayo is clearly the highest

Table 4.3 Unemployment indicators[1] (%)

	1990	2000
Unemployment rate in all sectors at national level	11	9
Unemployment rate in all sectors at city level, Bulawayo	16	17

Sources: 2000 figures are taken from the *1999 Indicator Monitoring Labour Force Survey*; those for 1990 are from the *1994 Indicator Monitoring Labour Force Survey* and *Labour Statistics* (CSO).
[1] Figures for the national level assume that communal farm workers and those in the informal economy are all employed.

in the country. Focusing on Bulawayo as a case study thus addresses a real problem in practice.

The unemployment rates are based on current employment, not usual employment. Current employment refers to what the survey respondents were doing in the seven days prior to the day of the survey (restricted definition), while usual employment relates to the last 12 months prior to the day of the survey (broad definition). The current employment figures may be affected by temporary jobs and seasonal working, depending on when in the year the survey is conducted. CSO (2004: 13) notes that 1992 recorded the highest unemployment rate at 22%, possibly due to the fact that the 1991–1992 drought affected agriculture, forestry and the fishing industry.

If communal farm workers and own account workers are considered as employed, the level of unemployment is very low (6.9% at the national level). This contrasts sharply with the 80% unemployment figures used particularly in media circles and by NGOs (see for example Rory Carroll, *The Guardian*, 2006: 21; Sachikonye, 2006).

Figures for the construction sector at both the national and city levels are taken as similar to the urban unemployment rates generally. The figures for those seeking work in the construction sector are not available due to the lack of a records system for those seeking work. Employment agencies are close to non-existent and people rely on word of mouth (social capital),

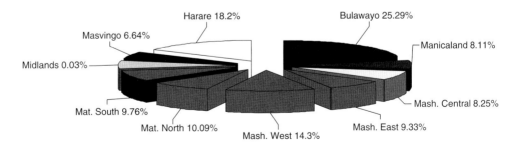

Figure 4.5 Unemployment by province in Zimbabwe, 2002
Source: Census 2000 National Report (CSO, 2002: 99), Government of Zimbabwe, Harare.

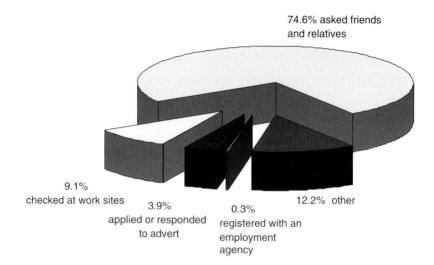

74.6% asked friends
and relatives

9.1%
checked at work sites

3.9%
applied or responded
to advert

0.3%
registered with an
employment
agency

12.2% other

Figure 4.6   Methods of job search for the unemployed in Zimbabwe
Source: 1999 Indicator Monitoring Labour Force Survey (CSO, 2000: 82), Government
of Zimbabwe, Harare.

as well as physical job searches in the industrial areas. The lack of formal
sector jobs also means that many unemployed are dissuaded from seeking
employment. According to both the 1994 and 1999 Indicator Monitoring
Labour Force Surveys, at least 75% of those seeking work asked friends
and relatives and less than 1% made use of employment agencies. These
job search patterns are summarized in Figure 4.6 and were confirmed in
field interviews. Tzircalle Brothers stated that they subcontract some of
their bricklaying, roofing, fitting and carpentry work to former employees
now working as either sole agents or small informal contractors.

   When jobs become available, no advertisements are posted. Instead, the
message is sent out through a vast network of contacts, which has been
progressively established since the 1970s. Similarly, informal builders in
Cowdray and Pumula South identified networks of friends, family and
former clients (in Bulawayo, Botswana and South Africa) as sources of
information on new jobs. The implications of this are that it would be inappropriate and inaccurate to attempt to estimate statistical indicators of
volumes of work, shift work, bonded work and so on. What can be said
is that the volume of available work has declined since 1996–1998, following buoyant years since the 1980s. Construction firms and tradesmen have
relocated to neighbouring countries, moved to other businesses or closed
down completely. When jobs become available, remaining tradesmen are
able to charge very high prices, even in the informal economy.

Low wages

The low wage rate is measured by the number of employed persons earning less than half the median wage for the given year or period. Table 4.4

Table 4.4   Low wage rate indicator (%)

	1990	2000
Low wage rate in all sectors at the national level	n/a	25

Source: 2000 figures are taken from the *1999 Indicator Monitoring Labour Force Survey* (CSO).

shows that in general about one third of the working population earns below half the median income. This method of presenting low wage rates is at variance with standard practice in Zimbabwe where the CSO uses average earnings based on the arithmetic mean. Table 4.5 aims to give the pattern of average earnings at the national level, a pattern that applies to all sectors of the economy and at all spatial levels.

Table 4.5 also shows that before the advent of Economic Structural Adjustment Programme (ESAP) in 1991, average earnings were very high in real terms but have not improved since the rapid decline in the early 1990s. The exchange rate of the Z$ to US$ remained stable during the 1980s (as low as 0.6093 in 1985 and 0.5147 in 1988), before rising sharply after 1991. While nominal wages have increased, the real wage declined rapidly, especially after 2001. This decline was associated with the liberalization of the economy, which included removal of wage controls (see Bond & Manyanya, 2002). Until the early 1990s, the difference between real and nominal wages was small but this has increased significantly since the mid-1990s. As depicted in Figure 4.7, real wages have shrunk rapidly, particularly in the last decade.

Wages alone are no longer enough. As discussed elsewhere, there are other inflation-busting packages that have significant positive impact on the welfare of workers, for example provision of transport and meals at the

Table 4.5   Average earnings per person per month and exchange rate during 1990–2001 for the national economy

Year	Z$ per unit of US$	Average earnings/person/month	
		Z$	US$
1990	0.3793	586.64	1546.64
1991	5.0511	683.71	135.36
1992	5.4815	779.90	142.28
1993	6.9350	891.65	128.57
1994	8.3871	1046.98	124.83
1995	9.3109	1266.17	135.99
1996	10.8389	1685.18	155.47
1997	18.6081	2387.73	128.32
1998	37.3692	3224.81	86.30
1999	38.1388	4649.84	121.92
2000	55.0660	8409.59	152.72

Source: CSO (2004: 74) *Zimbabwe Labour Statistics*, Central Statistical Office, Harare.

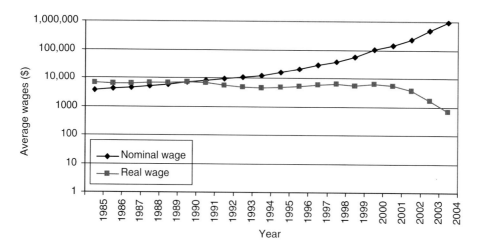

Figure 4.7   Nominal and real wages (1985–2004)
Source: Provided by the Labour and Economic Development Research Institute (LEDRIZ), Harare, March 2006.

workplace. The negotiation of wages is done through the NEC – a process in which Bulawayo City has no role, except when it comes to negotiating with those workers it employs.

Hours of work

For this study hours of work are calculated by measuring the proportion of wage earners working more than 45 hours a week at each given period. Forty-five hours is the cut-off point used by the Central Statistical Office in labour force surveys. This is in line with section 18 (2) of the SI 244 of 1999 (Collective Bargaining Agreement) for the construction industry, which stipulates that subject to exemptions, ordinary hours of work for all employees shall not exceed 44 hours per week (GoZ, 1999). Table 4.6 is based on adjustments of CSO data representing the proportion of workers doing more than 45 hours a week, with the assumption that the national figures apply to the construction sector. The most accurate data on this was compiled for the 1999 survey, which revealed that 44% of the working-age population worked over 45 hours per week, while 38% worked more than 48 hours per week[11]. Although hours of work for women in the wage sector were recorded at 35.3%, compared to men at 51.7%, women's hours of work are much higher when both domestic and informal economy work are considered.

Table 4.6   Hours of work

	1990	2000
Hours of work in all sectors at the national level	n/a	44

Source: CSO (2000) *1999 Indicator Monitoring Labour Force Survey.*

> **Box 4.2   More than hours of work**
>
> At its peak and even as recently as 2002/2003, Belmont Construction (Pvt) Ltd employed between 500 and 600 workers at a time. With punitive interest rates and inflation after 2003, like many other firms, Belmont could not afford to continue operating. It has retained only a skeletal staff, stopped construction work and moved into equipment hire. To support its few workers, it provides one meal a day on site as well as transport for employees not working at central sites.
>
> Interview with Edmund Zerf, Director of Belmont Construction, 11 April 2006.

With the collapse of the formal employment economy and the rapid growth of informal or self-employment, several changes have taken place. Employers and employees have agreed to reduce the number of working hours or work longer hours without overtime pay (or with no pay at all) in order to maintain operations and avoid closing down firms. Thus, in some cases hours of work have decreased while in others they have increased. The CSO (2000: 80) reports that up to 95% of workers preferred to work an extra two hours without being paid overtime in order to keep their jobs. On the basis of hours worked in the formal sector and extra informal work done in the evening and at weekends, it can be assumed that over 50% of workers now work more than 50 hours a week across all industries.

The cost of going to work or keeping operations going has increased since 2000. This has forced many construction companies to stop operating and has sent employees into the informal economy or self-employment, where they are likely to work longer hours than usual. The figures in Table 4.6 are estimates based on the assumption that since 2000, workers have had to work longer hours to maintain their minimum wage. See also Box 4.2.

Health and safety at work

The major source of data on health and safety at work is compiled by the National Social Security Authority (NSSA) and is based largely on submissions from employers, employer organizations, field inspections and new workers' compensation claims made under the 1990 Workers' Compensation Scheme, Statutory Instrument 68. NSSA statistics pertain to injuries and fatalities that occur in a particular year (for preventive purposes), although year of occurrence is hard to define for occupational diseases (NSSA, 2000: 5). The Central Statistical Office (CSO) also depends on NSSA statistics for its own publications. The records give details on reported occupational injuries by sector and region according to type of injury.

Table 4.7 Health and safety at work

	1996	2000
Days lost due to injury in the workplace per 1000 wage earners in all sectors at the national level	53.3	28.7
Days lost due to injury in the workplace per 1000 wage earners in all sectors in Bulawayo	6.7	47.5
Days lost due to injury in the workplace per 1000 wage earners in the construction sector at the national level	57.3	26.0
Days lost due to injury in the workplace per 1000 wage earners in the construction sector in Bulawayo	111.0	66.8

Sources: Compiled from NSSA figures April 2006 and *CSO Indicator Monitoring Labour Force Survey* and *Labour Statistics* (1994, 1999).

Neither the NSSA nor the CSO have any interest in collecting data on the number of hours of work lost to the economy due to occupational injuries or death. The NSSA only came into existence in 1994 and started collecting comprehensive statistics in 1995. Therefore, for Zimbabwe a comprehensive picture on safety at work can only be gleaned for the past decade or so. Key indicators from the data include total injuries per year, incidence rate and frequency rate, calculated per thousand insured labour force. The insured labour force is the number of workers registered with the NSSA for social security purposes. This is compulsory for all workers, except peasant agricultural workers, domestic servants and those in the informal sector. So theoretically, all workers in the formal sector are insured workers. The NSSA figures do not include civil servants.

Table 4.7 provides a summary of the number of hours lost to the economy due to occupational injuries, providing both a conservative and high estimate. How these figures were calculated and why the pattern of improving conditions may actually be the opposite of what happens in the economy needs further explanation. The input needed for this are figures on reported injuries, total number of the insured labour force and incidence rates. The occupational injuries figures were used together with the number of insured labour force to calculate the incidence rate, that is, the number of occupational injuries per thousand workers. The incidence rate was multiplied by three (the minimum number of days away from work for an injury to be recorded) to give a low estimate of the days lost to the economy.

NSSA figures for the 1995–1998 period show that since the mid-1990s, there have been a growing number of occupational injuries in the workplace. Any injury that does not result in absence from work or less than three days away from work is not recorded for the purposes of NSSA monitoring. This means that many other injuries in the workplace never get recorded. The NSSA (2000: 6) states that reported occupational injuries underestimate the extent of the problem due to the fact that:

Table 4.8   Reported occupational injuries in the building and construction sector

	1996	2000	2005
Occupational injuries in all sectors at the national level	20,330	14,507	8290
Occupational injuries in all sectors at city level, Bulawayo	447	3949	2391
Occupational injuries in the construction sector at national level	1578	926	366
Occupational injuries in the construction sector at the city level, Bulawayo	414	353	135

Source: National Social Security Authority (NSSA), April 2006.

- Occupational injuries to the self-employed are excluded, given that most workers in the informal economy are not covered under the workers compensation scheme.
- Cases occurring away from the workplace or on a journey to and from work are not included.

The NSSA also runs the Bulawayo Rehabilitation Centre for workers injured in the workplace. Workers are detained for 45.5 days on average at the Bulawayo Rehabilitation Centre. Clearly, such workers are those with serious injuries and are not the majority. According to key informants, there is underreporting of injuries at work and the true figure could be as much as five times the reported figure. Key informants also claim that with the foreign currency shortages, sourcing spare parts and equipment has become a serious problem, forcing companies to recycle old components and taking other risky measures that compromise safety in the workplace.

The reported occupational injuries in Table 4.8 can be used together with the number of insured labour force in Table 4.9 to calculate the incidence rate, that is, the number of occupational injuries per thousand workers (Table 4.10).

Recent legal changes on health and safety at work
As provided in the Factories and Works Act (1976) (GoZ, 1976) employers are legally bound to provide NSSA officers with access to companies and workplaces and to forward to NSSA records of occupational injuries. NSSA officers are authorized to issue prohibition orders to any offending employers or on sites where standards are not met.

Table 4.9   Insured labour force and the construction sector

	1996	2000	2005
Insured workers in all sectors at the national level	1,143,324	1,517,657	1,744,267
Insured workers in all sectors at city level, Bulawayo	199,335	249,287	285,545
Insured workers in the construction sector at national level	82,662	106,954	114,900
Insured workers in the construction sector at the city level, Bulawayo	11,194	15,845	17,526

Source: National Social Security Authority (NSSA) data, April 2006 (see also *Annual Report, 2001*: 109).

Table 4.10   Incidence rate: occupational injuries in the building and construction sector

	1996	2000	2005
Incidence rate in all sectors at the national level	17.8	9.2	4.8
Incidence rate in all sectors at city level, Bulawayo	2.2	15.8	8.4
Incidence rate in the construction sector at national level	19.1	8.7	3.2
Incidence rate in the construction sector at the city level, Bulawayo	37.0	22.3	7.7

Sources: National Social Security Authority (NSSA) data and *Labour Survey Statistics 2006.*

In contrast, although worker representatives at the shop-floor level and ZCTU health and safety officers can inspect sites and make records of occupational injuries or diseases, they cannot issue prohibition orders and can only make recommendations to both employers and the NSSA. These recommendations can be ignored. See Box 4.3.

While basic metal products, the mining and quarrying sector, agriculture and forestry have the highest reported occupational injuries, building and construction is one of the top five, with high occupational accidents in relation to the number of workers. It must be stressed that there is no recording of workplace injury and disease statistics in the peasant and informal sectors. Only a focused time series study can shed light on incidence rates and hours lost to the economy in these sectors of the economy. The magnitude of occupational injuries in Zimbabwe remains hugely under reported.

4.3.2   Social security indicators

Public social security coverage

If only formally employed persons of the working age population are considered, there is 100% insurance against risks of injury at work,

---

Box 4.3   Statistics on the insured labour force

NSSA advises that until recently, insured labour force data was obtained from the Workers Compensation Insurance Fund (WCIF). This data has been inconsistent and inaccurate. Since the population covered by the National Pension Scheme (NPS) is ideally the same as that covered by the WCIF, the NSSA uses the NPS data on insured labour force, and this appears to be more consistent. This means that there has been an increase in the insured labour force by sector, so the incidence rates calculated based on this denominator tend to be lower than in previous years. This fact should be borne in mind when comparing rates (NSSA, 2000: 4).

Table 4.11   Public social security coverage rate[1]

	1990	2000
Social security coverage rate in all sectors at the national level	n/a	15

Sources: National Social Security Authority (NSSA) 2006 and the *Census 2002 National Report* (CSO, 2002).
[1] Legally, all formal workers must be covered by the Public Social Security System.

unemployment, sick leave, etc. This is the basis of the figures in Table 4.11. However, we consider that only 15% of the working-age population is in formal employment, so social security coverage at the national level is also 15%. There is no quantitative information to incorporate coverage of the informal sector.

For a more complete view of social security at the national level, more information is needed. Until 1994, Zimbabwe did not have a comprehensive public social security scheme. The Ministry of Public Service and Social Welfare operated the Workers' Compensation Scheme (WCIF). In 1994, the Government set up the National Social Security Authority (NSSA) to administer the WCIF in parallel with other forms of social security. This was largely the National Pension Scheme, or Pension and Other Benefits Scheme (NPS). Both the WCIF and the NPS cover all workers, except those in the civil service, security services, domestic workers, informal sector workers and peasant farmers. The economic decline since 2000 has delayed the setting up of new schemes to cover these workers and the planned National Health Scheme.

Since colonial days, there has been an assumption that the majority of black Zimbabwean workers and peasant farmers have rural communal land rights as their main form of social security. For this reason, the notions of labour rights, urban citizenship and land have been tightly intertwined. But as argued elsewhere, for many urban workers communal land rights are not about production only but other cultural and religious needs as well (Mbiba, 1999). Productive agricultural lands, and the supplies needed to make land productive, are in short supply or beyond the reach of most poor workers. In addition, a significant urban population now has no desire to access rural land rights, which the state maintains as a framework for social sanction and control (Mbiba, 2001).

Notwithstanding the above, we have to recognize the existence of communal land rights and other forms of traditional social security or social safety nets (GoZ, 2005c: 21) used by the majority of black Zimbabweans. These include access to land rights in rural areas, family networks and community support. The Government has tried to re-invigorate some of these through Zunde Ramambo – a food security scheme managed by traditional chiefs. While these may not be directly available to urban poor, it should be noted that black Zimbabweans are very mobile, with family members switching between urban and rural residence as part of survival and accumulation. The question of land (both rural and urban) is integrated into this complex web of relations.

Beyond the NPS and WCIF, there are private and occupational insurance and pension schemes at industry and sector levels, such as the mining industry, railways, electricity and local government. The NSSA National Pensions Scheme (NPS) was designed to be a basic system complemented with individual savings, personal insurances and company-level occupational pension schemes. From its inception in 1994, the insurable earnings under the NPS was set at a ceiling of $Z4000 and remained at that level until 2001 (NSSA Annual Report, 2001: 7), when it was raised to Z$7000. With hyperinflation in the country since the late 1990s, the value of this benefit has been severely eroded, even with the subsequent reviews that raised monthly payments to Z$252,000 by 2006 (*The Herald*, Wednesday 3 May 2006). Further reviews were anticipated, but would not make a difference in an inflationary environment.

Both the NPS and the WCIF are dependent on the cooperation of employers, who must register with the NSSA, collect contributions and forward these to the NSSA every month. There are problems in the implementation of all three of these steps. Zimbabwe has no single employer data base and this compounds monitoring problems for the NSSA. In 2001, amendments were made to the NSSA Act to give inspectors the power to enter and search premises, inspect employers' records and returns, and enforce collections and contributions.

NSSA offices are concentrated in main urban centres and people in remote or rural regions have to travel long distances to submit claims or make inquiries. This factor means that there are more workers who do not have adequate access to national social security.

The money collected by NSSA is not just to be paid out as benefits in the short term but is to be invested as well. For years, workers and labour unions have demanded that some of this be invested in urban housing for workers but NSSA has been slow to implement a comprehensive housing programme. The bulk of inflows to NSSA are channelled to investments such as equities, prescribed markets, money market and real estate (NSSA, 2001: 27). In turn, return from investments accounts for the bulk of inflows to NSSA, with employers and employees making an almost equal contribution.

Old age pension

The old-age pension coverage rate is calculated by measuring the proportion of people aged sixty-five or over without pension coverage at the given dates and sectors respectively. Figures for the population aged sixty-five years and over can be obtained from national census reports, while the pension coverage can be obtained from the NSSA. However, the NSSA enumeration framework does not correspond to the one used by the census office for practical reasons, leaving us with no credible data on pension coverage.

In the 2002 census, for example, the number of people 65 years and over was recorded as being 3.55% (215,842) for the nation as a whole and 2.54% (17,196) for the Bulawayo region, that is, 2.5% of Bulawayo's population was 65 years old or over. However, when it comes to the NSSA figures on pensions, the national figure was 183,184 and 26,933 for Bulawayo. The figure for those receiving pensions in Bulawayo is greater than the recorded census figure. This does not mean that everyone aged 65 years and over in Bulawayo has a pension. The discrepancy arises from the fact that the NSSA keeps all those whose pension addresses are in Bulawayo on its books, even if they no longer reside there. Many of these may have migrated back to rural areas, where the cost of living is lower, but they are still able to collect their pensions from banks or travel occasionally to Bulawayo.

In the formal sector at the national level, only about 15% of those aged 65 years and over are not covered by an old age pension. This figure is too low because it underestimates the number of those not covered, largely because the NSSA figures include a large proportion of pensioners (often white Zimbabweans) still on its records but now living outside Zimbabwe in South Africa, Australia, New Zealand or the UK. This scenario indicates that with the existing statistics, it may be very misleading to attempt a calculation of pension coverage by region and sector. A further point is that NSSA pension statistics do not cover the civil service and state security personnel (army, police, prisons and airforce). The director of state pensions refused to release any of these statistics 'for security reasons'. All civil servants are covered by a public sector pension.

4.3.3 Indicators of workers' rights

Legislation on workers' rights and working conditions

Through the pro-worker 1985 Labour Relations Act (No. 16), an improvement to the 1984 Employment Act, rights were conferred on workers in areas of job security and freedom of association for African workers in line with aspirations of the nationalist movement, which had contributed significantly to the independence movement. This pro-worker statute followed hundreds of labour protests and strikes between 1980 and 1985 because of high worker expectations soon after independence. While it provided for improved worker rights in an attempt to meet ILO standards, its main weakness was the failure to harmonize labour relations laws; it left out civil servants generally. It also excluded senior local authority workers, whose conditions of work would be governed by the Urban Councils Act.

The second phase of challenges came around the late 1980s and early 1990s with the introduction of the Economic Structural Adjustment Programme (ESAP). Even before the program was formally pronounced, stagnation had set in as the Government grappled with economic problems and sought dialogue with national and global capital. The latter argued that the 1985 Act was too worker-friendly, rigid and a disincentive for investors (e.g. in terms of minimum wages and price controls). The private sector also argued that the law did not allow employers to hire and fire workers and that it gave Government too much room to interfere in industrial employer-worker relations.

With ESAP in 1991, the Government basically gave in to private sector demands and came up with the 1992 Labour Relations Amendment Act (No. 17). As the passing of the Act became imminent, confrontation between Government and workers intensified, with historical protest marches in July 1992 in Harare. This marked the divorce between the labour movement and the Government in post-colonial Zimbabwe. The then Minister of Labour, John Nkomo, responded to the protests by retorting that workers and the ZCTU were neither Government nor a political party, but that if they wanted to become one then they were free to do so. Minister Nkomo was indirectly inviting ZCTU to form a political party[12] – an invitation repeated by President Mugabe a few years later. These sociolegal changes set the context for the detailed discussions which are outlined in the remainder of this chapter.

Wage inequality between genders

By the late 1980s, Zimbabwe women had made strong progress in many fields, particularly education and health care (GoZ/USAID, 1996). However, women have not fared well in the economy and remained underrepresented in formal employment and access to land. Commercial and industrial opportunities remain beyond the reach of the majority of women. In a detailed study of 18,800 low-income households participating in GoZ housing programs funded by USAID (GoZ/USAID, 1996), it was shown that only 6% of the beneficiary households were female headed, even though up to 20% of urban households were female headed in the 1980s. While the legal instruments have been changed to enhance women's advancement, many gender barriers associated with Zimbabwe's patriarchal society remain. Women face harassment and logistical problems when they seek to assert their rights.

Traditionally, women are responsible for significant aspects of the house construction process. They also constitute up to 70% of those working in the informal sector generally. However, when it comes to construction in urban areas, the technical nature of the process and the gender barriers make it difficult for women to participate. For example, during house construction, women are faced with delays in obtaining building

Table 4.12   Earnings inequalities between men and women of working age

	Women	Men
Work done (national level)	54.6%	76.9%
Income bracket ≤ Z$500	29.0%	19.1%
Earning ≥ Z$3000	7.7%	17.7%
Construction Sector		
No work done	43.3%	11.9%
Work done	56.7%	88.1%
Income bracket ≤ Z$500	15.6%	18.7%
Earning ≥ Z$3000	12.1%	17.7%
Median wage Z$ (national economy)	250 (250)[1]	620 (1999)[1]
Median wage Z$ (construction sector)	250 (874)[1]	1249 (1249)[1]

Source: figures modified from CSO (2000) *1999 Indicator Monitoring Labour Force Survey, Harare*, pp. 77–79.
[1] The figure in brackets is the median earnings calculated for those who had done work only, i.e. excluding the 'no work done' category.

materials from providers, especially during times of shortage when materials are very expensive. Where women have to travel long distances or queue to get materials, they have problems concerning childcare and other domestic responsibilities and are subject to sexual harassment (GoZ/USAID, 1996: iv).

In summary, the majority of women have remained outside the formal employment sector and their presence in cities was restricted by colonial apartheid laws until independence in 1980. Unfortunately, this perception of women from the colonial era remains; hence the limited progress, despite laws against discrimination on the basis of gender, race or religion. Table 4.12 is based on the 1999 labour force survey and shows 45.4% of women have never 'worked', compared to 23.1% of men. This refers to paid work and is misleading, given that women do a lot of domestic and community work that is neither recorded nor directly remunerated. In rural areas remuneration for agricultural work is usually recorded under the name of the male spouses. According to the CSO (2004), employment rates in 1982 were 71.4 for men, compared to 42.9 for women, changing to 60.6 and 36.7 in 1992, and 73.0 and 62.0 in 1999 respectively[13]. Table 4.12 also shows that nationally, more women earn low incomes than men, with 29% of women in the lowest income category of under $Z500, compared to 19.1% for men. At the top end, 17.7% of men earned incomes in the highest bracket of Z$3000 and above, compared to 7.7% for women. Published labour statistics for 1990–1995 are not available to allow a presentation of trends. However, the 1999 labour force survey does give a picture of the gender inequalities in the construction sector where the work patterns are similar to those at the national level.

However, in the construction sector the income gap between men and women is narrower than at the national level. We assume that the construction sector figures are similar to those for the city level, where no published figures are provided in labour force surveys. Based on the median incomes for 1999–2000, at the national level men earned as much as eight times more than women. In the construction sector men earned about one and half times more than women. The bulk of low-paid workers (mainly women and children) are in the agricultural sector, such that if this is factored out of the national figures, the gap between the national median wage for women and men would be closer to that for the construction sector. In summary, although the Labour Relations Act 5 (1) protects employees against discrimination on the grounds of race, sex, tribe, place of origin and political opinion, Table 4.12 shows that in practice inequalities persist.

Wage inequality based on workers' places of birth

Places of birth for workers range from rural areas and other cities within the country, as well as abroad. CSO (2004) states that economically active immigrants from outside Zimbabwe have declined from a peak of 1826 in 1985 to 1087 in 1990, 647 in 2000 and 394 in 2002. Between 1990 and 2000 the highest numbers of male immigrants were architects and engineers, while most female immigrants were teachers. These CSO figures do not cover illegal immigrants, most of whom would find work in the agriculture and informal sectors. Although labour force surveys and the censuses give figures on foreign-born residents, there are no published statistics on income differences between workers' places of birth. However, considering that incomes in the agriculture sector are lower than in the rest of the economy, it is tempting to assume that workers in the construction sector in Bulawayo earn more than in their rural places of birth. This may not be the case, however, if we were to give a monetary value to all rural work and associated benefits.

Child labour

Prior to the 1999 Indicator Monitoring Labour Force Survey, information on child labour was scarce (CSO, 2004: 56). Children are defined as those less than fifteen years of age (CSO, 2001: 45), although for working children data, labour statistics CSO (2004) define children as those of five to seventeen years of age. Children do participate in active work, particularly in the agricultural, forestry and fishing sectors (87% of all working children in 1999), and the service sectors (10% of all working children in 1999).

   The CSO (2004: 56) indicates that in 1992, 3% of the 1,457,000 working children aged ten to fourteen years were largely unpaid family workers. The proportion working in the construction sector is negligible, at about 1.7% of all working children in 1999 (CSO, 2004: 59). For the whole economy in 1999, the CSO (2000: 51) reported that 8% of the 4,226,693 working children aged five to seventeen were employed largely as unpaid family workers (5%). These figures on child labour illustrate once again the inconsistency in definitions and frameworks used in official documents, in this case figures for children switch from ten to fourteen, to five to seventeen, to five to fourteen.

4.3.4   *Indicators of social dialogue*

Legislation on social dialogue

In terms of the Labour Relations Act (1996), all workers have the right to belong to a union and to form and participate in workers' committees. This section pays specific attention to the conditions of social dialogue and the legislative changes that have taken place since the 1980s. Strictly speaking, Bulawayo City has played a marginal role in this area, largely due to the legal and institutional framework that defines relations in this area of decent work. It is central government, employers, unions and the National Employment Council (NEC) that are central to deliberations on this issue where contradictions remain. Although the legal framework and conditions on paper have improved significantly since the 1980s, in practice some of these gains in the labour sphere are compromised by recent economic decline and overriding security laws.

Key informants agreed unanimously that legal and statutory provisions for workers' rights and social dialogue have improved immensely over the last decade and are better than before. However, in practice there are short-comings and often reversals in areas relating to civil servants, the informal sector, contract or casual labourers (who make up to 95% of workers in some firms in the construction sector), domestic servants and women. The key constraint is the declining economy. This has heightened political tensions in the country, leading to an atmosphere of a 'state of emergency' where, in the 'interests of state security', the Government has introduced measures reversing gains in social dialogue and workers rights. Key informants cited the implementation of POSA and AIIPA as the most damaging, even though on paper these statutes may appear beneficial to the nation.

At independence in 1980, the country had to transform from the inherited dual legal system to a unitary one and do away with racially inspired labour laws such as the Masters and Servants Act, under which African workers were not considered as employees. Key informants reflected with nostalgia on the early days in the 1980s when the ruling ZANU (PF) and its key members, such as then Minister of Labour, Kumbirai Kangai, with his philosophy of 'one industry one union', championed the cause of workers and harmonized the fragmented unions into a formidable Zimbabwe Congress of Trade Unions (ZCTU).

A Bulawayo-based trade unionist considers that the 1992 Amendment Act caused splintering of workers' activities, increased or created bureaucracy, disempowered trade unions and was a marked return to labour colonialism; a 'return to Master and Servants' through the back door. Through the SI of 1992, cited as the Retrenchment Regulations, (replacing SI 4040 of 1990) and in pursuit of ESAP policy, firms could now retrench workers more easily (the dreaded *chigumura*[14], as opposed to the situation in early 1986. The right to strike was curtailed, making it illegal for workers to go on strike without permission; and workers are of the view that permission is impossible to get. Furthermore, export promotion zones (EPZ), the EPZ Act (Chap. 56) exclusively and explicitly prohibited the application of the Labour Act in these zones. Bulawayo has one such zone, which remains undeveloped.

The massive deregulation of the economy – leading to company closures (e.g. Cone Textiles, National Blankets in Bulawayo) and massive retrenchments – impoverished workers and reduced their bargaining power at the shop-floor level. There was a rise in contract working. At the national economy level, the budget deficit along with both foreign and domestic debt increased. Both Government and workers found themselves under pressure and, while the Government relaxed laws for employers, it toughened laws for workers, who were left with little room for manoeuvre. Strikes and public demonstrations became the order of the day, including the landmark countrywide strike by civil servants in August 1996 and City Hall demonstrations in Bulawayo in 1997. Workers staged a series of job stay-aways from 1997–2000, activities that paralyzed industry and effectively reasserted the labour movement as a critical player in the national political economy at a time when civil organizations and other traditional allies of the ruling party (such as war veterans and students) were also challenging ZANU (PF) legitimacy and domination.

In line with ILO provisions, the Zimbabwe Government introduced tripartite negotiating in 1998 to bring together Government, workers and employers in a bid to create a harmonious economic environment in the country. Workers are represented by the Zimbabwe Congress of Trade Unions (ZCTU) although the political climate has seen efforts to put up a competitor union. At industry level, and in line with the Labour Relations Act, workers are represented by their Zimbabwe Construction and Allied Trades Workers Union (ZCATWU). Employers have to set up and register a National Economic Council and Collective Bargaining Agreement, both lodged with the Government. Employers and employees agree on wages and working conditions via the NEC and the CBA process. Employers are represented by members from the larger firms in the Construction Industry Federation of Zimbabwe and those from the Zimbabwe Building and Construction Association (ZBCA). All have regional offices in Bulawayo.

Table 4.13   Union density rate

	1992	2000
Union density all sectors at the national level	16	13

Source: Union representatives.
NB Union numbers include only those affiliated with the National Zimbabwe Congress of Trade Unions at the time of recording.

Union density rate

According to figures from the Zimbabwe Congress of Trade Unions (ZCTU) Information Department, total union membership was about 200,000 in 1990. This declined to 141,560 in 1995 and rose again to 165,012 in 2000 and 228,430 in 2005. This rise is also reflected in the construction sector, with membership at 3000 in 1998, up to 3700 in 2000 and 4800 in 2005. At the Bulawayo City level, membership rose from 500 to 750 and 1200 respectively[15]. Union membership and CSO figures on employed workers were used to calculate union coverage and density indicators, given in Table 4.13. However, these should be considered as very conservative. First, the original figures only cover returns from unions affiliated with the ZCTU based on paid-up membership[16]. This membership tends to be very high in an election year when unionists mobilize potential voters, as opposed to other years. Second, the original figures on total employees by industry sector at national level include civil servants not affiliated with the ZCTU. If civil servants were excluded from the employee totals, the overall union density would be much higher.

Schiphorst (2001: 229) asserts that, with the exception of the mining industry, trade unions in Zimbabwe are geographically organized and do not have a structural presence at the shop-floor level. Unions' capacity to organize and take collective job action has been restricted both before and after independence, although since 1985 the Labour Relations Act has incorporated the right to strike, except for those sectors considered 'essential services'. Union membership appears to have taken a fall during the mid-1990s, when many workers lost their jobs due to retrenchments as part of the economic structural adjustment policy. More precisely, 14,000 workers lost their jobs in 1993 alone (CSO, 2004: 26).

Collective bargaining

While the continuous collective bargaining process seems to be going well, agreements are not always easy to achieve and workers do go on strike. There are limits, however, on how much workers can use this collective action to express dissatisfaction. Without permission from the Government and the police and where action is deemed to be a political or security threat, strikes and demonstrations will be prohibited, often forcefully.

When agreement failed during the 2004 negotiations between the council and its workers, council workers resolved to demonstrate at City Hall, Bulawayo. Declared unlawful by the police, the demonstration was forcefully disbanded with reported injuries and deaths as a result (ZUCWU, 2006: 3). Labour unions allege that AIPPA and POSA make it possible to restrict social dialogue activities that would ordinarily be permissible under the Labour Relations Act.

At the regional level, the construction workers' union in Bulawayo is bound by agreements reached at the national level through both the TNF and CBA process. It is worth noting that when the Government raised fuel prices unilaterally in 2003, labour (ZCTU) felt the Government was not treating it with respect as a social partner and the national TNF collapsed. The dialogue resumed in early 2005. Other areas of national governance in which the ZCTU would like a greater say include the National AIDS Council (NAC) and the NSSA, given the contribution that workers make to these organizations. Unhappy with the management of NSSA, labour withdrew its representative on the NSSA board and the situation remained unchanged two years later.

Both the NEC for the construction sector and the workers union in Bulawayo are poorly equipped and technical expertise is overstretched, leaving no room for strategic work. Neither office saw any need or space for the local authority in promoting conditions of work and argued, as did the local authority, that conditions are set nationally and the only issues at the local level are about implementation. An area of concern for both the NEC and the local authority is the employment of unregistered contractors. Contractors may be registered as legitimate companies with the NEC but not with the contractor/employer organizations, ZBCA and CIFOZ, which set and monitor professional standards among their members. Ideally, the local authority should give contracts only to those members registered with both the NEC and ZBCA or CIFOZ – this appears not to happen all the time.

While the NEC is supposed to police, and has authority to charge unregistered builders and contractors, its ability to do so is limited by manpower. Field interviews with bricklayers revealed that informal builders previously registered with the NEC have defaulted but can still get construction jobs. The NEC has no resources to inspect sites, while contractors have devised ways of avoiding detection.

4.3.5   Workers' rights and social dialogue in Zimbabwe

The Government's response to worker agitation in the 1990s included passing the 1998 SI 368 (Anti Stay-away Regulations), the 2002 Labour Relations Amendment No. 17, the 2003 Amendment 7 and the 2001 Criminal Penalties Amendment Act. The latter criminalized labour misconducts. For example, it made it possible to prosecute those calling for a strike

(section 104 and section 107) and enabled employers to bring workers before a court of law and to demand they pay compensation for loss of production as a result of strike action, including lawful strike action. Other
key operational legislation on working conditions includes the SI 244 of
1999, as supported by the Labour Act 28: 01, No. 17 of 2002 and No. 7 of
2005. Section 3 of Act No. 7 of 2005 brings together the public and private
sectors, as well as those working in the export processing zones, previously excluded from this act. Salaries and other conditions are negotiated
through a continuous process as per the collective bargaining agreements
in the industry.

However, Mucheche (2005b: 3) explains that even with the 2005 amendments, workers in the armed forces, prisons and police remain excluded.
He argues that civil servants are deprived of full rights to collective bargaining, the right to strike and access to proper and efficient dispute
resolution mechanisms. Civil servants can still be unfairly transferred
without notice or consultation, leading to the separation of spouses; a separation that increases the risks of HIV/AIDS among heterosexual couples
in Zimbabwe. Yet, as confirmed by the Supreme Court, civil servants have
a right to be consulted before transfer[17].

Workers in the construction sector, through the Construction and Allied
Trades Workers Union and the ZCTU, consider that their continued struggles from the 1990s have led to some positive changes in 2003 in the legal provisions for workers countrywide, particularly in the context of the
2002 Labour Relations Amendment Act No. 17. According to ZCATWU
and ZCTU, these positive changes include:

- Protection from unfair dismissal: every worker has a right to not be
  unfairly dismissed.
- Right to strike: strikes are allowed in non-essential services and pick-
  eting is now allowed.
- Trade union superiority over workers' committees: where a registered
  trade union represents the interests of not less than half of the employees at the workplace where a workers' committee is to be established,
  every member of the workers committee shall be a member of the trade
  union concerned.
- Empowerment of workers council: that is, the managerial preroga-
  tive is diluted through employee participation in decision making. An
  employer can consult a workers' council regarding proposals relating
  to restructuring of the workplace following technological change and
  work methods, product development plans, job grading and training
  and education schemes affecting employees, partial or total plant closures, mergers and transfers of ownership, the implementation of an
  employment code of conduct, the criteria for merit for pay increases
  or discretionary bonuses, and the retrenchment of voluntary or temporary workers.

- Casual and contract work: there is new employment security for contract workers. A contract of employment that does not specify its duration or date of termination, other than a contract for casual or seasonal work or for some specific service, shall be deemed a contract without limit of time.
- Sick leave: this has been extended from one month to six, of which three months are on full pay and the other three on half pay, before the contract of employment can be terminated.
- Vacation leave: paid vacation leave has been extended from an average of 18 days to 30 days each year. Thus, since the 1990s, leave days have gone from 1.5 days a month to an average of 2.5 days a month.
- Maternity leave is granted for 90 days on full pay (up from 45 days). The other conditions remain the same.
- Notice of termination of a contract of employment by either party is now three months for contracts without limit of time (permanent) and contracts for a period exceeding two years; two months for contracts for a period exceeding one year but less than two years; one month for contracts for a period exceeding six months but less than one year; and two weeks for contracts for a period of six months or less and casual or seasonal work.
- Special leave on full pay is given to an employee for the following reasons:
  - On the instruction of a medical practitioner because of contact with an infectious disease
  - When required to attend court in Zimbabwe as a witness
  - When required to attend a meeting as a delegate or office bearer of a registered trade union representing employees within the undertaking or industry in which the employee is employed
  - When detained for questioning by the police
  - On the death of a spouse, parent, child or legal dependant
  - On any justifiable compassionate grounds
  This special leave is limited to 12 days in a calendar year.
- Discrimination based on gender, HIV/AIDS status or disability is now prohibited.

However, as noted by several commentators, while positive, the legal provisions in these amendments fall short of the needs of some categories of workers and face several obstacles. The biggest enemy is inflation (ZUCWU, Bulawayo, 2006). Since 2000, legal provisions and employers have made efforts to provide extra support such as grants for burial or funerals, soap for washing uniforms or overalls, travel allowances or transport to and from construction sites and central collection points in the city, and the provision of meals at work, as reported by key informants and respondents in Bulawayo[18]. Workers remain concerned with deterioration in health and safety at work as the economy finds it difficult to mobilize foreign currency and purchase needed spare parts. Observations of buildings in both Harare and Bulawayo show that on average only one-quarter to one-fifth of the elevators will be working on any given day. Some of the lifts in older buildings require total replacement and this costs 18 to 21 billion dollars for one lift, plus another 21 million per month to service the lift[19].

Construction sector workers have further specific social security concerns regarding the 65 years age at which benefits can be paid. Yet, according to both workers and employers in Bulawayo, construction workers hardly survive beyond 45 years and, under current conditions, by the time any of them get to 65 the pension benefits will be insignificant.

According to the ZCATU, while the average week for construction workers is 44 hours, they need to have this reduced to 42, but working this out for guards and casual workers remains problematic. When workers are injured, they claim there are unfair practices in the determination of the degree of injury (and hence compensation) and in job security for injured and casual workers. Respondents suggested that professional assessment by doctors representing the workers should also be taken into account in addition to those made by NSSA-appointed doctors. While the rehabilitation centre in Bulawayo is a good facility, it may be useful to have more such facilities, at least another centre, perhaps in Harare.

The construction sector remains a 'out of bounds' for women, both as workers and employers, with only about 12 women currently members of CIFOZ/ZBCA. While maternity conditions have improved with the Labour Relations Amendments (2003), further changes are needed, such as work suits for women, instead of overalls, and separate toilets (together with all related facilities), even if there is only one woman at the construction site. The greatest scourge in the workplace for women in the construction industry is sexual harassment. Harassment needs to be defined in the construction context in order to protect women (F. Mugabe). Women would welcome support in this area from others and the sharing of global experiences on ways of improving both regulations and management practice. Discrimination on the basis of pregnancy remains an issue in the sector and, despite the general legislative improvements on maternity and sick leave, Zimbabwe remains behind its neighbour South Africa in this area (Mucheche, 2005a).

4.3.6   Synthesis: decent work indicators in Bulawayo

Table 4.14 presents the decent work indicators for Zimbabwe and Bulawayo according to the four key components. The right hand column of this table shows trends towards (positive) or away from (negative) decent work for each of the indicators in all sectors and in the construction sector at the national and local levels.

Table 4.14   Bulawayo decent work indicators

Employment dimension				
Unemployment rate				
		1990	2000	Trend towards DW
National level	All sectors	11%	9%	+ve
Local level	All sectors	16%	17%	−ve
Low wage rate				
		1990	2000	Trend towards DW
National level	All sectors	n/a	25%	
	Construction	n/a	33%	
Hours of work				
		1990	2000	Trend towards DW
National level	All sectors	n/a	44%	
Social security dimension				
Public social security coverage				
		1990	2000	Trend towards DW
National level	All sectors	n/a	46%	
Old age pension				
		1990	2002	Trend towards DW
National level	All sectors	n/a	3.55%	
Local level	All sectors	n/a	2.54%	
Workers' rights dimension				
Wage inequality between genders				
		1990	2000	Trend towards DW
National level	All sectors			
	Construction			
Child labour				
		1992 (10–14)	1999 (5–17)	Trend towards DW
National level	All sectors	3%	8%	−ve
	Construction	n/a	1.7%	
Social dialogue dimension				
Union density rate				
		1992	2000	Trend towards DW
National level	All sectors	16%	13%	−ve

4.4   Decent work in Bulawayo: initiatives and evidence

4.4.1   *Equality and the indigenization policy in the construction sector*

At the time independence was gained in 1980, the construction industry was dominated and controlled by members of the Construction Industry Federation of Zimbabwe (CIFOZ) whose purpose was to serve and preserve the interests of large white dominated construction firms. The indigenous workers were excluded from this organisation not only through racial structures but also as a result of skewed access to technology, financial resources and expertise. The new majority Government was keen to reverse this disadvantage, but it took time to bring change to the construction sector. Around 1985, a group of indigenous contractors came together to form the Zimbabwe Building Contractors Association (ZBCA), which grew to be the largest single representative group for indigenous small and medium size contractors. Membership in 2006 was estimated at about 500[20]. It lobbied the Government to implement affirmative actions in support of marginalized workers in the construction sector.

Although many groups campaigned for better policy on indigenization and affirmative action economy-wide, the breakthrough in the construction sector only came in the early 1990s at the instigation of the World Bank. In its work in Zimbabwe, the World Bank realized the unsustainable nature of the inequalities in the construction sector and urged the Government to address the marginalization of indigenous agents. The outcome of this was the Government of Zimbabwe Treasury Circular No. 2 in 1993. The fact that the policy instrument on construction came from the Ministry of Finance and not the Ministry of Construction further illustrates the hand of the World Bank in the change (Sibanda, 2004). The incentives offered by Bulawayo City, as described in the section on urban development, should also be considered in light of this indigenization.

Treasury Circular No. 2 made it mandatory for all members of CIFOZ to subcontract between 7.5% and 15% of project contract value to indigenous members of the ZBCA, that all Government and parastatal projects award at least 10% of project work to ZBCA members and that all tenders of less than three million Zimbabwe dollars be awarded to ZBCA members only (GoZ, 1993). It was envisaged that the policy would be reviewed with the involvement of both CIFOZ and ZBCA and be mindful of ZBCA's capacity to absorb all such work. Towards the end of the 1990s, further indigenization policies were initiated in the economy, such as a policy framework for indigenization of the Zimbabwe economy[21]. For the purposes of the 1993 affirmative action policies in the construction sector, an indigenous building contractor was considered as one properly registered with both ZBCA and the Ministry of Public Construction and National Housing.

It is within this context that local authorities have implemented programmes to support local indigenous contractors. Without affirmative action, indigenous contractors were left to bid for the difficult and remote contracts that CIFOZ members would not be interested in. While not all ZBCA members have made progress, a review of G.G. Hardware and Construction (Pvt) Ltd clearly showed a slow diversification from small public sector projects in the 1980s and early 1990s to larger private sector jobs in the late 1990s and after 2000. With greater experience and more resources bases, G.G. Hardware moved from small projects of 15 to 30 housing units in Bulawayo's high density suburbs around 1991/1993 to bigger private sector projects of up to 200 housing units each, such as the Mimosa Mining Company projects in 2002 (G.G. Hardware and Construction, 2006).

Over the years, CIFOZ and ZBCA have operated in competition and debated the definition of 'indigenous Zimbabwean contractor'. Within the changing political and economic environment, there now seems to be agreement that the two bodies should merge into one and that this single organization will be more able to withstand globalization pressures. The experience of the past two decades will be used to maintain regulations that allow smaller and formerly marginalized groups to win reasonable construction projects. While all key informants were of the view that the merger is imminent, there was no clarity as to how inclusion mechanisms would be enforced. This is an issue worth monitoring.

4.4.2 Managing centre local relations

We concur with Mutizwa-Mangiza (1991) that the fortunes of a local authority (and that of Bulawayo City in particular) depend precariously on the fate of the economy and the political relations it maintains with central government. The relationship with local residents and business can also be added to this. Our study has shown that, comparatively, Bulawayo City has retained its high degree of political and administrative autonomy. It has earned great respect from central government and works well in partnership with business, workers and residents. Bulawayo City leaders expressed confidence that despite prevailing economic challenges and discordant party politics, Bulawayo's autonomy would endure, although Sachikonye (2006: 16) considers otherwise. The economy has nonetheless become one of the most difficult challenges facing Bulawayo. Dimensions of this problem are often beyond the control of city authorities. It is this context that has shaped Bulawayo's best practices of the past and calls for new approaches in the future.

This chapter has presented a series of indicators and conditions regarding the four components of decent work, followed by detailed discussions of legal changes and the frameworks for social dialogue. Among the conclusions, is the fact that while there have been improvements in legal provisions for decent work in comparison to previous decades, the practical experiences are not so positive for workers in government, for casual workers and for women. Bulawayo City's role in most areas of decent work is marginal, largely due to the institutional division of labour as prescribed by law. However, a range of best practices have been especially developed in areas of strategic planning and urban development, which all, directly and indirectly, create a climate conducive to improvements in decent work. We revisit some of the best practices.

4.4.3 Realistic strategic planning and citizen participation

Bulawayo has used existing planning legislation to implement planning processes that involve and respond to community and business needs. These strategic plans have provided certainty for investors and residents alike. The master planning process and outcomes in the early 1980s, as well as the recent ones at the turn of the millennium, have shown Bulawayo City to be a leader among other local authorities in the country. Not only has it introduced regulations, it has also been innovative by accommodating a range of informal sector activities, such as tuck shops, phone shops and urban agriculture. Bulawayo has a drier climate than other areas to the north-east of the country. While other local authorities have continued with prohibitive measures against urban agriculture, Bulawayo has recognized the activity, and put a promotion program in place that tackles poverty and environmental challenges in a way compatible with the long-term strategic needs of the city. The Netherlands-based RUAF Foundation and SNV are key international partners in this endeavour.

The master planning process has utilized local experts, in contrast with cities elsewhere in Africa that still depend on international consultants. Public consultation was central to strategic planning and continues locally, making use of the Citizens' Charter with organizations such as Bulawayo United Residents Association, Bulawayo Public Transport Association, Bulawayo Affirmative Action Group, churches, NGOs, the Zimbabwe National Chamber of Commerce and the Government of Zimbabwe. Similar consultations also take place annually at various stages of the budget cycle. While the process does not equate to full citizen empowerment, it does indicate that Bulawayo values the contributions of stakeholders in its administrative area and seeks to work in partnership with all of them.

It is recommended that the dialogue between Bulawayo City and the tripartite partners continues to seek solutions to the water problems and encourage business investments that take advantage of Bulawayo's hinterland and its proximity to markets in South Africa.

4.4.4 The role of Bulawayo in promoting employment creation

We have shown that since the 1980s, Bulawayo City has forged partnerships with the private sector, Government and donors (e.g. World Bank, USAID) to implement infrastructure and urban development programs. Transparency has been key to this partnership in that Bulawayo has put up clear indicators and incentives that enable all stakeholders to take part. These incentives were utilized to promote the indigenization program, although the gender dimension appears to be missing and needs to be promoted more actively in the years ahead.

As part of the indigenization program in a bid to create employment, Bulawayo has promoted training programs and incubator schemes. The Kelvin North incubator project is probably the most innovative enterprise promotion exercise. As part of their project, the council built factory shells which were allocated to beneficiaries on five-year leases. It was hoped that beneficiaries would start small industrial production activities and at the end of the five-year period would have grown bigger and be able to move to new 'open market' premises. The shells were to be advertised and new tenants brought in at the end of the five-year period. At the time of the survey in March/April 2006, almost all of the shells at Kelvin North were occupied and although carpentry enterprises were the most dominant, metal work (e.g. door and window frames), battery recycling and textiles were also some successful enterprises.

While key informants viewed Kelvin North as a successful project, Bulawayo City officials conceded that there was need for a thorough review of the program to identify the origins and destinations of the beneficiaries. Without a detailed survey and analysis of administrative records, it was not possible to tell what proportion of the Kelvin beneficiaries were from the school-leavers training program or retrenched from formal industry. At the same time, the policy objective that beneficiaries 'grow' out of the incubators appears not to have been met. Monitoring and enforcing the five-year leases has been minimal and allegations of misuse have been made by key informants at the site.

Of the 46 incubator shells at Kelvin North 1, about 46% were operating under sub-letting arrangements where the operator had no lease with the local authority, while 85% of the registered leases had expired in 2004. At Kelvin North Phase 2, leases for 15 of the 16 incubators had expired in 2005 and seven of the operators were running under sub-letting arrangements. There is certainly significant activity taking place at these project sites, but whether they remain within the expectations of council needs to be reviewed in detail.

In light of the fact that the incubators shells have been in operation for a long time, there is need for a comprehensive evaluation to see if the original policy objectives need to be revisited. Senior administrators have conceded this point and promised to look into obtaining resources so that these can be made available in the short term. The ILO could contribute financially to this evaluation.

4.4.5   Bulawayo and the promotion of cooperatives

The formation and operation of cooperatives is currently guided by the Cooperative Societies Act (1990 as revised) administered by the Register of Cooperatives in the Ministry of National Affairs, Employment Creation and Cooperatives. It is within this framework that the Kelvin North Training Centre was initiated around 1992–1993. It has created no less than 12 building cooperatives with a total membership of 120 and many more labourers employed on a casual basis. With ongoing economic hardships, only one cooperative remained in 2006. They build hundreds of houses but officials are not able to provide exact figures. There were also three cooperatives dealing in metal work (whose peak membership was 60), but only two remained in 2006. Other cooperatives initiated for crafts, agriculture, arts and drama have also found it difficult to remain in business since the economic difficulties of the post-2000 period.

4.5   Decent work: evidence, obstacles and potential

4.5.1   *Decent work and development in Zimbabwe*

As previously stated, the prevailing socio-economic context of Zimbabwe poses serious challenges for the promotion of decent work. The economy declined by 32% between 1999 and 2005 (ZCTU, 2005b: 1) and business confidence has also declined continuously (Intermarket Research, 2004: 36–40). Over the past six years, Zimbabwe has been the only SADC country where investment has fallen to extremely low levels and both growth and employment creation have been negative (CB Richard Ellis, 2006; ILO, 2005c: 6; Intermarket Research, 2004;). In this context, the goal of economic recovery and employment creation becomes the highest priority relative to other decent work goals; whether the jobs created can be considered as decent work or not becomes a secondary issue. This is why workers offer to work extra hours without pay – simply to keep their jobs.

The previous chapters have demonstrated that while an empirical outline of the decent work indicators in Zimbabwe can be compiled, the bulk of the statistics will only be estimates showing trends in workers' conditions. Whether data is collected from national institutions, employer organizations, trade unions, enterprises or individual workers has a strong bearing on the accuracy and validity of the information. Although the decent work concept has a recent history, demands for improvement in working conditions are not new and recall the struggle of Zimbabwean workers from colonial days to the present. However, few in Zimbabwe expressed familiarity with the term 'decent work' and even worker representatives in the construction sector were ignorant of it. Thus, the challenge of decent work is not only conceptual but methodological. The key question is how can it be measured? In regards to Zimbabwe, this includes finding the right institutions to facilitate the dissemination and implementation of the agenda at the local level while locating this agenda in national priorities.

The ongoing ILO/SAMAT program to promote decent work has crystallized around the recently launched Zimbabwe Decent Work Country Program 2006–2007 (ZDWCP). This has been largely a national affair in which local authorities have had little or no involvement; decent work is not a pertinent operational concept and the ILO is not visible at the local level. Broad consultations for the ZDWCP involved tripartite partners: Government, the Employers Federation of Zimbabwe (EMCOZ) and ZCTU. In this national program the objective was to integrate decent work into the country's development plans, with Zimbabwe prioritizing Millennium Development Goal 1 on poverty reduction, Millennium

Development Goal 3 on social protection and reduction of the impact of HIV/AIDS in the workplace, and Millennium Development Goal 6 on upholding and strengthening social dialogue and tripartite consultation.

For the decent work concept to succeed at the local level, it is imperative that the ILO broadens its work in Zimbabwe beyond 'traditional stakeholders' and expands to work with more micro-level partners. It is important to note the macro nature of the ZDWP and hence the need to translate it to local level actions. HIV/AIDS in the workplace appears to have been a key area of activity for labour unions for a long time now and can be combined with job creation in the informal sector to provide a package that responds to the country's needs.

Recommendation

We recommend that work to reduce HIV/AIDS in the workplace not only be strengthened but also broadened and used as a launch pad to consider other aspects of decent work, especially gender, occupational accidents and post-accident care. To be more visible, the ILO should take a multi-pronged approach at the local level; working with NECs, local authority unions and strategic teams, as well as collectively with all of them. Second, promoting dialogue continues debates on broader issues of employment creation and poverty alleviation. Bulawayo is already well positioned to be receptive, considering its work in the urban agriculture sector and the best practice in industrial incubators.

4.5.2   *Methodological and conceptual considerations*

This separates the population to expose the different categories and highlight the labour dimensions. However, unlike the 1999 Indicator Monitoring Labour Force Surveys (CSO, 2000: 45), the 2002 census (CSO, 2002: 83) does not display disaggregation to show the communal farmers component within the economically active population. Ideally, the economically active population should be divided into the unemployed, the communal area farmers, commercial farm workers, the rest of formal employment and informal sector employment.

The inclusion or exclusion of peasants and the informal sector has also varied, with most international organizations and NGOs putting these in the unemployed category. As we saw earlier in this chapter in Section 4.3, there is no consistency in the methods used for data collection over time and across institutions. There are variations even within the CSO. The 1997 Inter-census Demographic Survey (ICDS) analyzed activities for the ten to fourteen-year age group, while the national census of 1992 did not. We also noted that the CSO definition of children switches between ten to fourteen years, five to seventeen years and also to five to fourteen years old. See Figure 4.8.

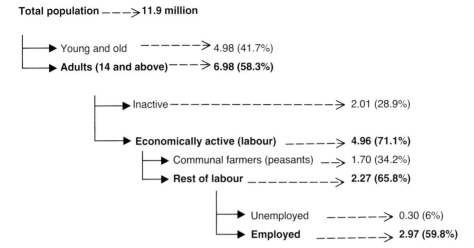

Figure 4.8   Population structure and labour (1999–2000)
Source: Modified from *1999 Indicator Monitoring Labour Force Survey* (CSO, 2000: 45).

In the national census for 1992 the broad definition of unemployment was used, yet all other censuses and surveys before and after have used the strict definition. In the strict definition, unemployed persons are those of age 15 and over who, during the period of reference, were without work and were actively seeking employment. The indicators of actively 'seeking employment' used in surveys do not capture the social capital-based methods utilized by most unemployed people. In the broad definition, the difference is that the criterion of 'actively looking for work' is excluded. Using the strict definition generally used by the Zimbabwe Government leads to statistically low levels of unemployment. The reference period has also been used selectively, varying between seven days for the 1982 census and all the IM-LFS, and 12 months for the 1992 and 2002 census and the 1987 and 1997 ICDS. Thus, figures for unemployment rates vary depending on whether the strict or broad definition is used, whether a long or short reference period is used, and whether the peasants and the informal sector are defined as employed.

Political tension and economic crisis has made data sharing and research a very sensitive issue. Individuals and institutions feel so insecure that they are not prepared to release even what should be routine administrative and public information (e.g. NSSA office in Bulawayo). These perceptions make data collection more difficult and research less plausible. A major methodological problem is that statistics collected for different purposes and at different institutional levels either do not add up or tell only part of the story. It would be more fruitful if data collection for this decent work framework could be repeated in subsequent years and incorporated into the work of key institutions such as CSO, NSSA, NECs and ZCTU.

Recommendation

We recommend that the ILO strengthen its dialogue with CSO and NSSA regarding data collection and analysis format. This dialogue should target the periodic surveys done in particular by the CSO, with a view to incorporating decent work data frameworks into the design of the data collection and analysis.

There has to be an obvious benefit to all these organizations before they can adopt some of the given frameworks or indicators. NSSA is best placed and has shown willingness to improve its data compilation exercises and, in particular, to add the hours lost to the economy due to workplace accidents and injuries. There must also be pressure from constituencies such as the labour movements and NECs for such mainstreaming to gain ground. As part of the national decent work program in the short or medium term, it is necessary to strengthen NECs as partnership platforms for all stakeholders in promoting decent work.

Statistically, the empirical level of data collection is important; whether the research uses data from individual workers and their unions at shop floor level, whether data comes from regional and national representative organizations, or from national authorities and departments such as CSO and NSSA. This study uses data from all sources, although it is clear that a comprehensive shop-floor level study for the construction sector is needed.

4.5.3 Women's rights at work and health and safety in the construction sector

The labour movement and workers' current health and safety work has rightly focused on HIV/AIDS, with a view to using the workplace as a source of information on the epidemic and circulating information on best practices among workers in regards to protecting and working with those affected. This theme is also central to the recently unveiled Decent Work Country Programme, which has scope for further improvements at the local level.

Recommendation

First, the ZDWP priority could be the most appropriate framework to mainstream the sexual and labour rights of women in the workplace. Using the same platforms, specific issues affecting women in the construction sector can be given a more central role with a view to changing men's behaviour as well as management practices. In particular, clarity on definitions of harassment, based on women's experiences in the construction sector, would need to be more explicit, together with the penalties against violations. This is not an issue for the construction sector alone but for the entire Zimbabwe labour force, although those in the construction sector could benefit from their counterparts' experiences in other parts of the world.

Literature and key informants in Bulawayo show that there have been improvements in the legal provisions that benefit women in areas such as maternity leave. However, in practice both workers and officials at the local level are not well conversant with how this applies and often have to refer to the Ministry for clarification. A key aspect of gender equality is to evaluate the extent to which the new legal provisions are understood and implemented. Achievements can then be used as a process to build confidence in the new legal provisions among all those involved.

A second point is that occupational health and accidents at work remain key issues that need to be addressed. Underreporting both in the formal and informal sector is of particular concern. The existing systems are over-stretched but could be effective with more local level inter-institutional dialogue and harmonization. The proposed decent work audit would provide a construction project based framework to revisit issues of health at work and give the local authority an entry point to promote conditions of workers, even if these are not its employees.

4.5.4   The informal economy: opportunities in Bulawayo

The need to promote the informal sector as a generator of employment was recognized as far back as the Growth with Equity policy document and the First National Transitional Development Plan (1982–1985). Through SI 216 of 1994, enacted in terms of the Regional Town and Country Planning Act, the Government relaxed controls. This gave local authorities room to permit employment creation activities in residential areas. Activities such as tailoring, bookmaking, wood and stone carving, car repairs, carpentry, shoe repair and tin smithing are some of the activities that blossomed. This book has given best practice details in the provision of industrial incubators in Bulawayo (e.g. Kelvin North). These complemented a broader Home Industries policy promoted throughout the country in the late 1980s and 1990s in which city councils provided serviced land for commercial SME activities close to low-income residential areas. The policy environment also led to informal activities even in the CBD, especially following the ESAP years in the 1990s.

In this chapter we have pointed to the rapid growth of the informal sector since 1990 and noted the shift from manufacturing to trading and services in both the formal and informal sectors. Kanyenze et al. (2003) further noted that in the informal sector over the same period, women have been squeezed out of the manufacturing sectors and into services and trading as well as from large operational units to increasingly small ones. However, in general, despite growth in informal sector jobs and internal changes, no detailed studies on decent work are available. The Training and Research

Support Centre (TARSC) has flagged occupational health issues (Loewenson, 2001b), but these remain a key concern, especially in the context of HIV/AIDS[22]. Thus, any future decent work initiatives in the informal sector must take gender and occupational health into consideration, as with the economy in general.

In May to July 2005, the Government launched Operation Murambatsvina (OM) which destroyed a significant number of these informal enterprises and settlements throughout the country, as a method of dealing with perceived economic and environmental problems in urban areas (GoZ, 2005c). NGOs and the 'international community' condemned the policy and have continued to express concern about the negative impacts on people's livelihoods[23]. There is still debate on the reasoning behind the operation (Potts, 2006; Sachikonye, 2006). Unfortunately, the few detailed studies on the impact of this policy have not included Bulawayo (Sachikonye, 2006). However, the OM appears to be a reversal of the Government's SI of 1994. Court rulings in Bulawayo regarding this policy have directed all parties to move forward through dialogue and Bulawayo City is pursuing this vigorously.

As a consequence of OM, there is an ongoing vigorous and visible countrywide reorganization and relocation of small enterprises and informal sector activities. This has generated new opportunities to support employment creation and decent work dialogue. There is a good opportunity in creating jobs through the informal sector, not only to evaluate the impacts of Operation Murabatsvina but also as a method by which the ILO can make itself visible at the local level and address urgent economic national needs.

Recommendation

We recommend that in Bulawayo, the ILO encourage stakeholders to revisit the role of the informal sector, making use of experiences from the home industries and industrial incubator programs. The dialogue on job creation should seek to move beyond Murambatsvina by meeting the needs of the poor while upholding the standards expected by Government, into which the decent work standards can be mainstreamed.

We recommend that the ILO support detailed research and documentation into the dynamics, employment and economic impacts of informal sector construction in Bulawayo including the role of the diaspora *injiva*. This is a feasible short-term project which would act as a focal point for broader policy and program creation for the informal sector in Bulawayo post-Murambatsvina. The project can have direct impact on the Reserve Bank of Zimbabwe's Homelink project.

Sharing of knowledge and experience with decent work and the informal sector should cascade to cities in Zimbabwe in the short term (one to two years) and to cities in the SADC in the medium term (two to five years) and others in Africa such as Nairobi (Kenya) and Accra (Ghana) where programs to reposition the role of the informal sector are also ongoing.

4.5.5 Future activities: Bulawayo's procurement dividend and the decent work audit

As a local authority, Bulawayo has expertise and experience in using environmental and financial audit schemes (Bulawayo City, 2000). It has also demonstrated commitment to partnership with the private sector, Government, workers and its residents. Donors have also worked with Bulawayo City in the past, and even in the current climate of economic decline, some international organizations have invested their resources to help with the city's urban poverty programme. Bulawayo City is not only a large employer in the city, it is also a consumer of resources and services and a purchaser and commissioner of extensive building and construction related work. It has the most potential to change people's welfare and to build stronger local economies and sustainable communities. It can and should use the above assets to respond to the challenge of decent work and in particular to help change attitudes and patterns of behaviour among all those, worker and corporate, operating within its area of jurisdiction.

Recommendation
We recommend that using its procurement services and the commissioning of construction works, Bulawayo City can put in place a decent work audit scheme that will act as a policy instrument to encourage and reward those organizations and institutions that strive to promote decent work. The indicators to the audit would be in line with priorities of workers and the ZDWCP: progress on health and safety at the workplace, on reducing the impact of HIV/AIDS and gender and social dialogue – largely assessed by the workers themselves.

There is synergy between what the local authority does, and the patterns of behaviour and welfare of people in the city. Within the climate of partnership that prevails between Bulawayo City and other actors, the decent work audit scheme would operate in tandem with the environmental management audits and EIAs. The key is that it is not Bulawayo City alone that will do the monitoring but all the stakeholders within the local partnership. The NECs, NSSA, employer and employee unions would be encouraged to incorporate these audits into their negotiating forums. As part of the tendering process, Bulawayo City would reward and award contracts to those making progress towards meeting decent work conditions for their workforce.

There are currently obstacles that would work against this recommendation. There is weak institutional capacity at the level of construction industry and worker's organization in Bulawayo. Consequently, capacity building and investment in manpower in these institutions is needed as part of the broader programme of decent work and to ensure that the decent work audits make a difference in practice.

Employers are likely to view such a decent work audit as simply another unnecessary burden added to the challenges of operating in a hyperinflationary environment. It may therefore be more effective to promote the scheme more forcefully only when the economy starts to recover. A further challenge is that informal sector workers and employers who are outside the existing administrative regimes would find it difficult to incorporate the audit scheme.

4.5.6   Obstacles and potentials for decent work promotion in Bulawayo

Decent work promotion has never been a direct policy area for the Bulawayo City Council, largely due to the way its responsibilities are defined in the statutes. However, all its activities in service delivery and translation of Government policy into local programs, as well as its relations with its own employees, are indirectly relevant to decent work.

Employment creation and setting the conditions that enable investment are key areas of potential. While the city has done a lot to provide infrastructure, land and incentives to invest, its marketing strategy has not taken a vigorous regional approach to capturing the potentials offered by existing transport infrastructure, population flows and proximity to Botswana and South Africa. It must also begin to make better use of the National University of Science and Technology.

Obstacles to decent work in the area of employment creation include the negative image of Bulawayo as both water-scarce and indifferent, as witnessed during the drought and water crisis in 1991–1992. There is a global awareness regarding the need to deal with climate change and the future of 'desert cities'. Bulawayo needs to be put in this bracket of cities and have support mobilized to deal with its water problems. It suffers from the negative impacts of the ESAP that led to decline in manufacturing.

Centre local relations are also essential. While direct interference by Government is not at levels similar to cities like Harare and Mutare, Bulawayo suffers from contradictions because it is perceived as a perennial home of the ruling party's political opponents; first to ZAPU until 1987 and then MDC from 1999. Central government's neglect in making timely investments in the water capacity of the city is often placed amongst these political contradictions (Zaaijer, 1998).

The potential to play a role in areas such as social dialogue, health and safety at work, equality and so on all depend on the room for manoeuvre provided in the legislation and centre local relations. In the future, Bulawayo can only take a policy and regulatory role in these areas in partnership with employee and employer organizations.

Appendix 4.1   Inventory of public land under private hands (developers) in Bulawayo City 1990–2006

	Developer	Status	Stands allocated and area of the city	Year allo-cated	Purchase/transactions	Development status 1999	Development status 2006
1	ZIGEU	Ind.	500 Cowdray Park 3	1995	$1/4$ payment made in 1993	No development yet. To commence development (1999), signed agreement	Stands paid for in full. House construction about 90% complete. Roads, sewer, water about 60% complete
2	Ministry of local government and national housing	GoZ	500 Cowdray Park 3	1996	No payment to date (1999)	MLGNH requested a quotation of stand prices and servicing costs (Dec. 1998). No feedback to date	Stands fully paid up and building plans approved. 419 stands serviced and houses built. Working with council to complete balance
3	Project management and Turnkey Project (PMTP)		5000 Cowdray Park 1 and portion of Cowdray Park Phase 2. 2000 Cowdray Park 3 Total 7079 stands	1996	Partial payment for Phase 2 No payment made for Phase 3	Completed Phase 1. Currently working on portion of Phase 2 No progress on Phase 3	Phase 1 fully paid up and building plans approved. 6079 stands fully developed and occupied. Balance of stands swapped with stands in Pumula South currently under development
4	Zimbabwe Building Society (ZBS)	Ind.	2000 stands – Cowdray Park 2	1996	No payment made so far	Negotiating with contractors and finalising funding	Withdrew due to capacity problems and council is reallocating the stands
5	Alpha Construction	Ind.	500 Phase 2 Cowdray Park	1996		Presently constructing roads. However, development has started	242 stands fully paid for, developed and occupied. 90 stands at various stages of development, 16 fully developed but not occupied, 152 stands not serviced or developed. Taken to court for double allocation of stands and building across stand boundaries

(Continued)

Appendix 4.1   *(Continued)*

	Developer	Status	Stands allocated and area of the city	Year allocated	Purchase/transactions	Development status 1999	Development status 2006
6	ZIMAL	Ind.	10000 Phase 3 to 4 Cowdray Park	1996	Paid quarter (deposit) of the purchase price and signed agreement	Have submitted development programme and commenced development	390 stands fully serviced but not developed, 223 stands fully developed allocated and occupied, 207 stands being developed; balance of stands now being developed by Council and Government under the 'Hlalani Kuhle' housing development scheme
7	ZIRUS	Ind.	2494 Pumula South	1997/8	No payment made so far	Developer still sourcing funds and resources to develop	Council has repossessed all stands for lack of development. To be reallocated
8	Glenkara Belmont		1000 stands and 699 stands Pelandaba South	1997	No payment made	No progress. Land partially serviced with outfall sewer. Slow progress on land purchase negotiation for part of the land to be occupied by the scheme	Only 25% of purchase price paid. No agreement with council and no building plans have been approved
9	NSSA	GoZ	Selborne Park Town Houses Sites A1	1998	No payment made	Unserviced land – no development to date	Council offer has been withdrawn due to lack of progress
10	Hound Investments		Selborne Park Town Houses Sites A2	1993	No payment made	Unserviced, no development to date	Council offer has been withdrawn due to lack of progress
11	Black Consult	Ind.	Selborne Park Town Houses Site B	1997	No payment made	No development. No response to reminders	Council offer has been withdrawn due to lack of progress
12	A.G. Georgiades		Selborne Park Town Houses Site C	1997	No payment made	No development. Asked for indefinite extension for payment and turned down	Council offer has been withdrawn due to lack of progress

13	Siyakha (Pvt) Ltd	Ind.	Selborne Park Town Houses Site C	1997	No payment made	No development. Have asked for six months extension to start project	Council offer has been withdrawn due to lack of progress
14	Siyakha (Pvt) Ltd	Ind.	Selborne Park Town Houses Site D	1997	No payment made	No development. Developer has requested six months extension to start project	Council offer has been withdrawn due to lack of progress
15	Siyakha (Pvt) Ltd	Ind.	Selborne Park Town Houses Site E	1997	No payment made	No development. Developer has requested six months extension to start work	Council offer has been withdrawn due to lack of progress
16	Project Management and Turnkey Projects		Selborne Park Town Houses Site F	1997	No payment made	No development. Working on building plans	Council offer has been withdrawn due to lack of progress

Source: Data up to 1999 from Bulawayo City (1999), Outstanding Development of Virgin Land Offered to Private Contractors for Residential Development, Town Lands and Planning Report dated 15 February 1999 to council on 3 March 1999. Data for 2000–2006 obtained from key informant interviews and field observations in April 2006.

Key: Ind. = Indigenous contractor or developer; Non-ind. = Non-indigenous developer.

Appendix 4.2 City of Bulawayo: incentive table (1990s)

	1	2	3	4	5	6	7	8		9	10	11	12	13
		Land incentives						Guaranteed project approval time						
	Allocation	Price	Lease option	Water	Free land	Property tax holiday	Prof. services	Appli-cation	Plan	Plan fees	Incu-bation	Accomm-odation	Elect-ricity	Telecom
a Employs ≤ 100 people														
b 101–300 people		80%											A	A
c 301–500 people		75%											A	A
d 501–700 people		70%											A	A
e 701–900 people		60%											A	A
f ≥901 people	G	50%											A	A
Size of Investment (Z$ million)														
g ≤$5						1 Year		90 days	30 days	50%			A	A
h 5–10 million						2 Years		90 days	30 days	60%			A	A
i 11–30 million						2 Years		60 days	30 days	70%				
j 31–50 million						3 Years		60 days	30 days					
k 51–70 million						3 Years		60 days	30 days			A		
l 71–100 million						4 Years		60 days	30 days			A		
m ≥100 million	G					5 Years		30 days	15 days			A		
n New and modern technology			A											
o Local and indigenous person			A				A				A			
p Export oriented			A											
q Pioneers			A											

Key: G = Guaranteed; A = Available
Source: Bulawayo City (1994) *Incentives for Development*, Projects Committee Team Report 22 March.

Notes

Chapter opening image: For one US dollar a day, Grace cleans the windows of a building under construction on the seaside of Dar es Salaam, Tanzania. Photograph courtesy of Marcel Crozet, ILO.

(1)   The Kondozi Estate case is noteworthy. This was an enterprise which employed thousands of workers and conducted daily exports of horticultural produce to both South Africa and Europe, but the estate was seized and handed over to the Agricultural and Rural Development Authority (ARDA) in 2004. It is now derelict with land portions used to grow maize under the 'Operation Maguta', a food security project led by the Zimbabwe National Army.

(2)   See, for example, 'Chinese get ready to cash in on Zimbabwe building boom', *Financial Gazette*, Harare, 20 January 2005.

(3)   See for example, 'Construction industry appeals for support: little activity taking place countrywide. Established companies settling for small jobs', *Business Herald*, Harare, 27 January 2000 and '10,000 likely to lose their jobs in construction', *Business Herald*, Harare, 11 January 2001.

(4)   See for example, 'Closures impact on construction industry', *The Daily News*, Harare, 14 February 2003.

(5)   See for example, 'Chinese get ready to cash in on Zimbabwe Building Boom', *Financial Gazette*, Harare, 20 January 2005.

(6)   Amendments of the statutes have abolished the office of the Executive Mayor and reverted to the old system of ceremonial mayor with effect from mid-2008.

(7)   The role of ZINWA in urban areas was abolished with effect from 1 February 2009 and responsibilities returned to urban councils. Bulawayo City contested/resisted the role of ZINWA until it was abolished.

(8)   Elections in 2006 in Chitungwiza to elect mayor and councillors did not bring an end to the appointed commission. However, following national political dialogue in an attempt to reverse the country's crisis, the two major political parties decided to elect a non-executive mayor for Harare from outside the political parties. Thursday 3rd July, the 46-member Harare council chose Muchadeyi Masunda, a lawyer, to be non-executive mayor. This was preceded by amendments to the Urban Councils Act that scrapped the office of executive mayor.

(9)   Bulawayo hosted five of the eight largest textile manufacturers, such as Merspin, Cotton Printers, Merlin and Security Mills (Zaaijer, 1998: 23).

(10)   One US$ = Z$55.0660 at 2000.

(11)   This data is not available in the Labour Force Survey Reports using 45 hours as the cut-off point.

(12)   See Munhumeso vs Minister of Labour, Zimbabwe Law Reports.

(13)   Employment rate is the number of employed persons aged 15 years and above divided by the total population in that age group multiplied by one hundred.

(14)   *Chigumura* literally translated means 'vicious displacement' and was the term used by Africans to describe forced retrenchments in industry in colonial days.

(15)   Figures provided by Zimbabwe Construction and Allied Trades workers Union, Bulawayo.

(16)   ZCTU has 35 affiliates. There were about 33 in 1995.

(17)   See Taylor vs Ministry of Higher Education and Anor, 1996 (2) Zimbabwe Law Reports 772 (S).

(18)   During a visit to Belmont Construction, workers were having late tea with bread provided by the employer.

(19)   Estimates in Z$ as at April 2006 provided by Mrs A.N. Kwangwama, Head of Department, Rural and Urban Planning, University of Zimbabwe.

(20)   Five years earlier, the *Business Herald* put the figure at only 120. See 'Building Materials Up 56pc, says CSSO', *Business Herald*, Harare, 26 April 2001.

(21)   The fast-track land reform of 2000 and beyond is often referred to as a broader form of indigenization of the economy.

(22)   See 'Informal Sector Exposed', *The Herald*, Harare, Friday, 30 June 2006.

(23)   See Zimbabwe Report at http://unhabitat.org. See also a study on the impact of 'Operation Murambatsvina/Restore Order' in Zimbabwe, ActionAid International – Southern Africa Partnership Programme (SAPP-ZIMBABWE) in collaboration with Combined Harare Residents Association (CHRA) and Zimbabwe Peace Project (ZPP), August 2005.

5 Dar es Salaam

Jill Wells

5.1 Introduction

This chapter presents the case study of Dar es Salaam, Tanzania. It begins with an overview of the political context and an examination of economic development at both the national and local levels. Available information, including official statistics, has been analyzed for the years 1990/91 and 2000/01. Section 5.5 summarizes the key findings of the case study, highlights specific examples of good practice and recommends action needed to promote decent work. The final section presents some examples of efforts made by the city authorities to promote employment, along with the obstacles encountered during the implementation of this task.

5.1.1 *Background on Tanzania*

The United Republic of Tanzania lies just south of the equator, between Kenya and Mozambique, and is bounded on the east by the Indian Ocean. Tanganyika gained independence from Britain in 1961 and Tanzania was formed in 1964 by a loose federation between the mainland and the islands of Zanzibar, Pemba and Mafia. The port city of Dar es Salaam, with a magnificent natural harbour on the Indian Ocean, is the major commercial trade centre and the de facto capital.

Tanzania has more than 130 different tribes and almost as many languages. The decision to use Swahili as the national language was a significant factor in unifying the nation. On the mainland, 1% of the population consists of minorities (Asians, Arabs and Europeans), although in Zanzibar the population is much more mixed. There is a substantial refugee population in the west of the country; 444,000 from Burundi and 154,000 from the DRC.

The country has significant natural resources (diamonds, gold, natural gas, iron ore, coal and other minerals). Agriculture is the main economic activity for 80% of the population. This accounts for 43% of GDP and a substantial proportion of exports, comprising coffee, cashews and cotton. However, cultivated crops cover only 4% of the land area. With the prevalence of droughts and floods, it is questionable whether the environment is really suited to agricultural production. The problem is that there are few alternatives. Industry is underdeveloped, comprising only 17% of GDP. Mining has picked up in recent years, with an increase in the production of gold and other minerals, but employs very few people. After a period of stagnation in the 1980s, GDP started to grow in real terms from the middle of the 1990s and in recent years growth has been quite impressive. However, GDP per capita is estimated at less than US$300 and is unevenly distributed. In 1993 the bottom 10% received 2.8% of the national income and the top 10% received 30%. Fifty per cent of the population lives below the poverty line.

Life expectancy at birth fell from 50 in 1990 to 43 in 2002, but rose again to 51 in 2005[1]. Eleven per cent of the population is estimated to be infected with HIV/AIDS. The population is young, with 44% under 15. This implies both a high dependency ratio (made worse by the loss of the most productive workers due to AIDS) and a large number of new entrants coming into the labour force each year.

5.2 National, regional and local context

5.2.1 *National context*

From the end of the nineteenth century until the First World War, Tanganyika was a German colony. In 1918 German East Africa was divided and Tanganyika became a British protectorate under a mandate from the League of Nations. Zanzibar was already administered by the British. Tanganyika obtained its independence from the UK in 1961, whereas Zanzibar was not independent until 1963. In 1964 the two countries merged to form the United Republic of Tanzania.

Tanzania has an elected president, who is the head of state and government. Zanzibar has a separate president who is the head of government for internal matters. There are two elected national assemblies, one on the mainland and another in Zanzibar. Until 1995 Tanzania was a one-party state. From 1995 onwards, it has had multi-party elections but the main party, Chama Cha Mapinduzi, or CCM (Revolutionary Party), still dominates. At the last election, in December 2005, the CCM Presidential candidate, Jakaya Kikwete, was elected with 80% of the votes. However, in Zanzibar the population is much more divided and the opposition to CCM is much stronger. The last two elections were hotly contested and accompanied by violence, but CCM managed to maintain its hold on power by the narrowest of margins.

Political and economic development in Tanzania in the years since independence can be divided into three periods corresponding to the three phases of government. Each phase was under a different president and had its own unique characteristics.

First phase of government (1961–1985)

The first and longest phase under President Julius Nyerere, affectionately known as *Mwalimu* (teacher), stretched from independence in 1961 until 1985 when he resigned from the presidency. This was the period of consolidation of the nation and the building of Tanzania's unique brand of socialism under its single party (TANU, later CCM). In the early years of independence, Tanzania tried to attract foreign investment and build a market economy. This policy met with little success, as most of the foreign investment in the region was attracted to the more developed neighbouring country, Kenya. In this context the ruling party TANU began to rethink its development strategy. The eventual outcome was the Arusha Declaration of 1967, which established the egalitarian and self-reliant principles of Tanzanian development and set the country firmly on a socialist path.

Over the next few years, the Government embarked on a broad program of nationalization of plantations, industries and financial services. This was followed in 1972 by a government takeover of second homes and buildings. Most of those affected were from the Asian business community, many of whom decided to leave the country. The exodus of much of the business community, together with a ban on all forms of capitalist activity by members of the ruling party, effectively closed the door on the possibility of capitalist development.

The state subsequently embraced the task of developing Tanzania's fragile economy. In addition to controlling the enterprises it had nationalized, it also embarked on an ambitious program of industrialization. Many of the new industries were producing substitutes for goods which had previously been imported. The enterprises were large scale, capital intensive and, even at the time, did not seem very appropriate in the Tanzanian context. While some failed to get off the ground, others were able to survive because they were protected from competition by external tariffs and their monopoly in the domestic economy. Although the state owned at least 51% of the shares in these parastatals, they operated with some autonomy and many were managed by foreign companies. State-owned enterprises also took over the marketing of agricultural products and the state had a monopoly of international trade. According to Coulson (1982), the years after the Arusha Declaration saw a proliferation of public institutions, with parastatals or government corporations set up in almost every sector. In 1967 there were 64 parastatals, but by 1974 this number had grown to 139 and was still increasing (Coulson, 1982). By the 1980s there were 450 (Tripp, 1997).

The construction sector did not escape from state control, with the establishment of a state contracting company, Mwananchi Engineering and Construction Company (MECCO), in 1966 and a state consultancy a few years later. Construction units were also established in many of the ministries, local authorities and parastatals and much of the construction program was undertaken by direct labour, known locally as 'force account'. It was not until the 1990s that Tanzania seriously set out to develop a private construction industry.

Another important feature of Tanzania's socialist policy was the idea of village-based collective farming. Development of rural areas, where the majority of Tanzania's population lived and worked, assumed the highest priority in the Arusha Declaration. The policy for rural areas and the agricultural sector combined the idea of *ujamaa* (which Nyerere defined as 'familyhood') with that of the earlier village settlement policy to form a new policy for the development of *ujamaa* villages. The movement of the population into villages was seen as a prerequisite for the development of agricultural productivity and the provision of services to the rural population. A further motivation was the idea that communal working would reverse the trend towards class differentiation in the rural areas. This movement was supposed to be voluntary and it was, initially. However, after 1970 pressure began to be applied by over-zealous government and/or party officials. In 1973 the TANU Biannual Conference resolved that within three years the whole of the rural population should be living in villages, and by the end of 1975 almost everyone had moved (Coulson, 1982).

By the mid-1970s, the state dominated every aspect of economic activity and the party dominated the state. In 1975 the Government controlled 65%

Table 5.1   **Average annual percentage growth rates of GDP at constant (1976) prices**

	1980	1981	1982	1983	1984	1985	1986	1987	1988	1989	1990	1991	1992
GDP	3.0	−0.5	−0.6	−2.4	3.4	2.6	3.3	5.1	4.2	4.0	4.8	3.4	3.6
GDP per capita	−0.8	−3.5	−2.5	−5.3	0.2	1.6	0.8	0.7	1.2	1.7	0.9	0.8	2.9

Source: National Accounts of Tanzania, 1976–1992.

of wage employment, increasing to 73% in 1984 (Coulson, 1982). It regulated prices, wages, interest rates, internal and external trade and investment. However, when the economy was hit by a series of external shocks in the latter part of the 1970s, the state could not sustain this kind of economic expansion and became increasingly dependent on foreign aid and loans. External shocks included a rise in oil prices in 1973 and 1979, declining terms of trade for agricultural products, the break up of the East African community in 1977 and the war with Uganda in 1979. These external factors exacerbated the internal causes of the crisis. The most important of these was a drop in agricultural production, particularly of export crops, as a result of poor prices and inefficiencies in state marketing. Agricultural production was also adversely affected by the nationalization of large estates and the village based program, as well as recurring drought (URT, 1993). Industrial production also fell, despite massive capital inflows. The overall effect of the decline in both agricultural and industrial production was a rapidly increasing balance of payment deficit and a severe and chronic foreign exchange shortage.

These developments are reflected in the national account statistics. The data in Table 5.1 shows that Gross Domestic Product (GDP) actually declined in 1979, 1981, 1982 and 1983. Even when it was growing, this was barely sufficient to offset the increase in population. GDP per capita fell in five of the six years between 1979 and 1984. In 1992 it was still below the 1977 level. There is good cause to label the 1980s as the 'lost decade' for development.

The early 1980s were a time of acute shortages in every basic commodity, characterized by rationing and long queues to acquire provisions and meet other basic needs (Tripp, 1997). These shortages and permits fuelled corruption in the country. They also fuelled inflation and led to further falls in production as spare parts and fuel became unobtainable and crops could not be collected but were left rotting at the farm gates. By 1984, shortages of foreign exchange, rapid inflation, and fiscal and external imbalances brought the economy almost to a halt. Tanzania had reached a point where the economic inefficiencies and distortions resulting from state intervention in markets outweighed the political gains (Tripp, 1997: 61). A change in direction was inevitable.

Before the end of the first phase of government some steps were taken to chart a new direction. A National Economic Stabilization Policy was introduced in 1980 and Tanzania's own Structural Adjustment Program in 1982. But the steps were modest and differences between the Tanzanian Government and the IMF prevented an agreement until 1986, after Nyerere had resigned as President.

Second phase of government (1986–1995)

The second phase of government, under President Mwinyi, is associated with structural adjustment, the opening up of the economy and the privatization of state institutions. The period witnessed two major changes, from a planned to a capitalist economy and from a single to a multi-party state.

In 1986, the new Government launched a three-year Economic Recovery Program (ERP). This program included adjustment to exchange rates, an increase in foreign exchange allocation, higher producer prices and the lifting of price controls. However it could not resolve the balance of payments problem. Under pressure from bilateral donors, who threatened to withdraw aid, agreement was reached with the IMF in 1986 for a Structural Adjustment Facility (1987–1990). This was followed by a second three-year adjustment program (Social and Economic Action Program, 1989–1991) and the extension of the IMF Facility to 1994.

The conditions for IMF support included rapid currency devaluation, constraints on wage increases and cuts in public services. Not surprisingly, this brought out deep divisions within the party and between the party and the Government. The divisions crystallized over the causes of Tanzania's crisis – whether it was the result of external forces alone or internal policies. There were related disagreements over the course of future policy, trade liberalization and the role of the private sector. But the Government under President Mwinyi prevailed in following a liberalizing course.

The results of the structural adjustment policies were mixed. Agricultural output rose by an average of 4.9% a year between 1986 and 1993. Manufacturing output also rose but continued to be constrained by erratic power supply and poor infrastructure. GDP began to grow, but this growth was insufficient to balance the increasing population, and GDP per capita hardly improved. In 1993 it actually declined, as shown in Table 5.2. On the social front, cost sharing in education and health services

Table 5.2 Average annual percentage growth rates of GDP at constant (1992) prices

	1993	1994	1995	1996	1997	1998	1999	2000	2001	2002
GDP	0.4	1.4	3.6	4.2	3.3	4.0	4.7	4.9	5.7	6.2
GDP per capita	−3.9	0	0	0	4.0	0	0	2.0	2.0	3.8

Sources: National Accounts, 2002; National Bureau of Statistics.

placed additional burdens on the poor and the primary school enrolment rate fell from 92% in 1977 to only 40% in 1994. External balances remained a problem: the current account deficit widened, external debt soared and dependence on donors increased dramatically. Donors were increasingly worried about corruption and in protest they suspended aid in 1994. Thus, although the economy showed signs of improvement, it was still very vulnerable.

Third phase of government (1995–2005)

The third phase government under President Benjamin Mkapa consolidated the economic gains of the early 1990s. GDP grew steadily from the mid-1990s and began to outpace population growth, leading to an increase in GDP per capita (see Table 5.2). The Mkapa years also saw a further opening of the economy to private sector participation. The period witnessed a dramatic decline in the power of the state in the economic sphere and a transformation of its role from direct intervention in the economy to that of facilitator for private enterprise.

Mkapa was elected in 1995 in Tanzania's first multi-party election. Although CCM is still the dominant party, its role is greatly diminished and civil society organizations are now encouraged. The Mkapa Government afforded high priority to fighting corruption, which had escalated in the Mwinyi years, as well as the promotion of good governance and, above all, the maintenance of sound economic policies. The fact that today Tanzania is a favourite of the World Bank and lauded as one of Africa's success stories is a measure of the success of the Mkapa years.

Fourth phase of government (2005–)

Tanzania entered its fourth phase of government with the election of the CCM candidate Jakaya Kikwete to the presidency in December 2005, with 80% of the vote. Kikwete has vowed to continue with the sound economic policies of the previous government and to fight corruption on all fronts. At the time of writing the mood in Tanzania was optimistic. However the country remains desperately poor. GDP per capita is still estimated to be less than US$300.

5.2.2   *Economy, employment and the construction sector*

The structure of any economy can be described by analyzing the contribution of the various economic sectors to GDP, as well as by the structure of imports and exports. Alternatively, the focus can be on the activities engaged in by the 'economically active' population. This will be the approach adopted in this section.

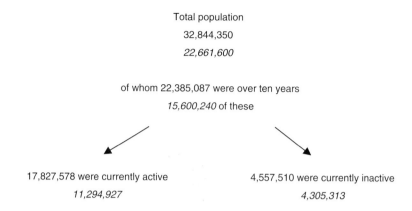

Total population

32,844,350

22,661,600

of whom 22,385,087 were over ten years

15,600,240 of these

17,827,578 were currently active	4,557,510 were currently inactive
11,294,927	*4,305,313*

Figure 5.1 Total and economically active population, 2000/01 and 1990/91 (figures in italics)
Source: Integrated Labour Force Survey, 2000/01 (URT, 2002).

Employment situation

The Integrated Labour Force Survey (ILFS) was carried out in 2000/01 as the second survey of its kind (URT, 2002). The first was carried out in 1990/91. The results of the two surveys are not always comparable but some similarities can be drawn. Both were household surveys covering only mainland Tanzania. The sample comprised 3660 households in urban areas and 8000 in rural areas, drawn from the National Master Sample. Everyone over the age of ten years in the selected households was included[2].

The total population and the numbers who were 'economically active' in 1990/91 and 2000/01 are shown in Figure 5.1. It can be seen that almost 18 million people were included in the 'Economically Active Population' (EAP) in 2000/01, compared with just over 11 million in 1990/91. This represents a 58% increase in the EAP between the two censuses. The increase in the total population during this period was only 43%, so a part of the increase in the EAP was due to a change in the definition of employment. The 2000/01 survey took the widest possible definition of the EAP to cover all those who were employed or available for work for at least one hour during the week prior to the survey. Economic activities included in the definition of employment were production for own use, unpaid family labour, housing services by owner occupiers and fetching water and firewood (included for the first time in 2000/01).

Data on the distribution of the employed population in 1990/91 and 2000/01 are shown in Tables 5.3 and 5.4 respectively. It can be seen that the vast majority of the population in both years (84% in 1990 and 81% in 2000) were working in traditional agriculture. The next largest is the informal sector. It can be seen that 9% of the working population in the country as a whole was employed in the informal sector in 2000/01. In the

Table 5.3   Distribution of employed population by main sector, 1990/91

	Government	Parastatal	Traditional agriculture	Informal	Other private	Total
National	319,455	180,767	9,115,932	955,647	317,404	10,889,205
National (%)	2.9%	1.7%	83.7%	8.7%	2.9%	99.9%
Urban	178,757	161,553	624,156	530,704	198,418	1,693,588
Urban (%)	10.5%	9.5%	36.8%	31.3%	11.7%	99.8%

Source: *Labour Force Survey 1990/91* (URT, 1993).

urban areas the figure was much higher at 33%. Although agriculture was still the main activity, accounting for 37% of the urban employed population, the informal sector came a close second. This is double the number working in 'other' private activities.

Comparison between the two tables shows a small increase in the proportion of the workforce in the informal sector and a small decrease in the proportion in traditional agriculture. These apparent changes may be due to changes in definition, however, together with the inclusion of household work in the definition of employment in 2000/01[3]. More significant is the fall in the proportion of the population employed by both the Government from (2.9% to 2%) parastatals (from 1.7% to 0.4%), together with the increase in the proportion working in 'other' private (from 2.9% to 4%) over the ten-year period. These changes are even more apparent in the urban areas, where the proportion working for government and parastatals combined fell from 20% to 8.4%, while those in 'other' private activities increased from 11.7% to 16%. The decline in public sector employment between 1990 and 2000, as well as the importance of the informal sector in both years, is seen more clearly when employment in the agricultural and household sectors (2000 only) are excluded. Results are shown in Figure 5.2. Whereas in 1990, 28.2% of the employed population outside agriculture was working in the public sector, by 2000 the proportion had fallen to 16%. This was a result of both cutbacks in the civil service and privatization of the parastatals. During the same period, the proportion employed in 'other' private sectors increased from 17.9% to 29%. However, it cannot be assumed that those retrenched from the Government or

Table 5.4   Distribution of employed population (standard definition) by main sector, 2000/01

	Government	Parastatal	Traditional agriculture	Informal	Other private	Household
National	344,839	78,270	13,694,935	1,439,848	756,046	600,867
National (%)	2%	0.4%	81%	9%	4%	3.6%
Urban	174,993	68,196	1,077,285	963,888	471,325	162,643
Urban (%)	6%	2.4%	37%	33%	16%	5.6%

Source: *Integrated Labour Force Survey 2000/01* (URT, 2002).

Distribution of employed population 1990/91

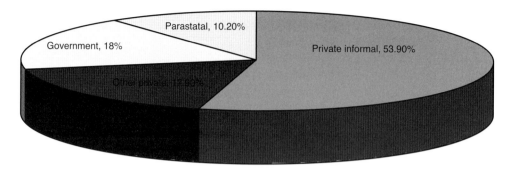

Distribution of employed population 2000/01

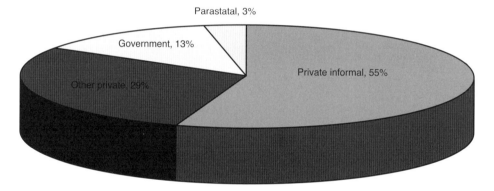

Figure 5.2   Distribution of employed population outside of agriculture, 1990/91 and 2000/01
Sources: Labour Force Survey, 1990/91 and *Integrated Labour Force Survey, 2000/01* (URT, 1993; 2002).

parastatals instantly found employment in the formal private sector, as 'other' private activities include employment in non-governmental organizations (NGOs), community-based organizations (CBOs) and cooperatives. Also, conditions of employment in the private sector may be very different from those previously enjoyed by public sector employees, an issue which will be returned to later in this chapter.

Employment in construction

Employment status is also recorded in the labour force survey by industry. The distribution of construction workers in mainland Tanzania by employment status in 1990/91 and 2000/01 is shown in Table 5.5. When assigning respondents to a particular industry, enumerators attempted to identify informal sector activities. A breakdown according to employment status

Table 5.5   **Distribution of employed construction workers (standard definition) by employment status and main sector, 1990/91 and 2000/01**

	Paid employees		Self-employed with employees		Self-employed without employees		Unpaid Family labour		Total	
1990/91	Number	%	Number	%	Number	%	Number	%	Number	%
Total	70,195	77%	7184	8%	13,026	14%	1244	1.3%	91,649	100%
Informal	—	—	—	—	—	—	—	—	—	52%
2000/01										
Total	55,413	37%	24,792	16%	66,943	44%	4,543	3%	151,690	100%
Formal	50,848	—	4230	—	3986	—	1441	—	60,504	—
Informal	4565	—	20,562	—	62,957	—	3102	—	91,186	60%

Sources: *Labour Force Survey, 1990/91*; *Integrated Labour Force Survey 2000/01* (URT, 1993; 2002).

and main sector (formal versus informal) is also shown in the table for 2000/01. The detailed breakdown was not published in 1990/91, so only the formal/informal breakdown of the total is given.

Table 5.5 shows that the estimated total number of construction workers in the country as a whole rose from 91,649 in 1990/91 to 151,690 in 2000/01. There was an apparent increase in the proportion working in the informal sector from 52% to 60%, but some of that increase is due to a change in definition. The most significant development in the ten-year period undoubtedly lies in the change in employment status. In 1990/91, 77% of the construction workforce was in paid employment, whilst 22% were self-employed. By 2000/01 this situation was reversed, with 60% self-employed and only 37% in paid employment.

Additional data on employees in the formal part of the industry is available from a survey of employment and earnings in the formal sector conducted in 2000/01 (NBS, 2004). This survey included all establishments in the public sector, all registered private establishments with more than fifty employees (profit and non-profit making) and a sample of those with five to fifty employees. The survey found a total of 818,860 paid employees in formal employment; 51% of these were employed in the private sector (compared to 27% in the previous survey of this kind conducted in 1984). Construction accounted for 7% of those formally employed in the private sector and 1% of those in the public sector. Altogether, 36,000 employees were found working in the formal construction sector, almost half of them (15,743) in Dar es Salaam.

The survey also distinguished between regular and casual employees. The former includes all permanent and temporary employees who have worked on a weekly or monthly basis for more than one month, while the latter includes all those receiving daily wages and employees who have not worked for the full month (June 2001). It was found that 85% of workers across all sectors were regulars and 15% casuals; whereas in the construction sector, casuals accounted for 15,051 out of a total of 36,935, or

41%. If the 4720 construction workers in the public sector are deducted from the total, casuals make up almost half (47%) of the 'formal' construction workforce.

A study undertaken in 2003 for the ILO of employment conditions on 11 large construction sites found an even higher proportion of casual employees, generally over 70% and sometimes reaching 95% (ILO, 2005b)[4]. This suggests that the proportion of casuals may actually be much higher than is captured in the employment and earnings survey. Seven of the eleven sites included in the survey were civil works in the rural areas and the other four were building projects in Dar es Salaam. The proportion of casuals on one of the building projects was lower than average, only around 50%, but this figure is misleading as it refers only to the staff of the main contractor and chief subcontractors, and they were not the main employers of labour. Most of the labour on the site was actually engaged for a specific task (rather than for a day or a month) through subcontractors. This is in fact the usual arrangement for building work in urban areas. Research in four towns in 1999–2000 (including Dar es Salaam) found that formal construction employs most of its workforce by out-sourcing to small enterprises working as labour contractors (Mlinga & Wells, 2002). Out of the fifty-five medium and large contractors interviewed for this publication 93% confirmed that they outsource work to small registered contractors and 82% said they subcontract to unregistered (informal) labour contractors. Almost all major building works are sublet on a labour only basis (Mlinga & Wells, 2002). Employment opportunities for informal construction workers in these small 'enterprises' (many of which are simply loose associations of workers or labour cooperatives) are limited and, in this sense, they may be more insecure than those directly employed as casuals (Jason, 2005).

Construction output

Output from the construction industry broadly follows that of the economy as a whole, but with wide fluctuations from year to year. When the economy is declining, construction output tends to decline at a faster rate (sometimes with a time-lag) with construction output constituting a falling proportion of GDP, and vice versa. This pattern is detectable in official data, as shown in Tables 5.6 and 5.7. From Table 5.6, it can be seen that

Table 5.6 Average annual percentage growth of GDP and construction output as a percentage of GDP, 1977–1992 at 1976 prices

	1980	1981	1982	1983	1984	1985	1986	1987	1988	1989	1990	1991	1992
GDP	3.0	−0.5	−0.6	−2.4	3.4	2.6	3.3	5.1	4.2	4.0	4.8	3.4	3.6
Construction (% GDP)	4.0	3.8	4.0	2.4	2.8	2.5	2.8	4.0	4.3	3.3	5.4	4.5	4.5

Source: National Accounts of Tanzania, 1976–1992.

Table 5.7   Average annual percentage growth of GDP and construction output at 1992 prices

	1993	1994	1995	1996	1997	1998	1999	2000	2001	2002
GDP	0.4	1.4	3.6	4.2	3.3	4.0	4.7	4.9	5.7	6.2
Constr. (% GDP)	4.6	4.6	3.78	3.91	4.09	4.32	4.49	4.64	4.77	4.99

Source: National Accounts, 2002, National Bureau of Statistics.

construction output stagnated at around 4% of GDP between 1976 and 1982, before falling to around 2.5% between 1983 and 1986. This was a few years behind the dramatic drop in GDP between 1980 and 1984. In 1983, construction output was just 2.4% of GDP, which is extremely low by international standards.

The opposite process can be observed from the data in Table 5.7. GDP was growing steadily year on year from 1997. Construction output was also growing steadily but at a faster rate, hence occupying a steadily increasing percentage of GDP. Official data on construction output could not be obtained beyond 2002, but it is believed still to be increasing at a faster rate than GDP. The World Bank reports that construction output rose by 11% in 2004 and GDP by only 6.7%. At the time of writing, Tanzania was experiencing a 'mini boom' in the construction sector.

5.3   Regional and local context

5.3.1   The evolution of local government

Under Nyerere's leadership, the Government saw the participation of the people as vital to the success of the nation's development plans. Since the beginning of the 1960s, the concept of 'development from below' has had considerable appeal (Kulaba, 1989). However, according to Tripp (1997), it soon became apparent that the main vehicle for participation was to be the ruling political party. Participation was not seen as a local, autonomous, grass-roots organization outside the party sphere. The party had a broad membership (2.5 million in 1988) but its structure was hierarchical. Every ten houses had a party cell so that directives from the party could filter down. Although it had been the intention in creating the ten-cell system that the views of the people should also filter back up to the leadership, this rarely happened in practice (Tripp, 1997).

The chance of popular participation in decision making was further eroded when in 1972 the national government abolished local government. Urban councils were effectively abolished from January 1974, with the management of urban areas integrated into the rural administrative system as parts of districts and regions. Dar es Salaam became a city region, divided into three districts (Ilala, Kinondoni and Temeke) and administered by regional and district secretariats appointed by, and responsible to, the central government. The policy was called 'decentralization' but, according to Coulson (1982), it could equally well have been called 'centralization', as it replaced elected councils at the local level with salaried central government officials. The 'decentralization' concerned staff not power.

There was also a quite deliberate bias in the allocation of resources during the period in favour of rural areas, with urban areas starved of funds (Coulson, 1982). In addition, the officials who were appointed to manage urban districts were agricultural experts who knew nothing about city planning. In the face of rapid urban growth, this bias led to a rapid deterioration in the standard of urban service and infrastructure (Kulaba, 1989). Only four years after the abolition of urban councils, there was a public outcry over the deterioration of services. The absence of authority to enforce development controls had also led to widespread flouting of urban by-laws and building regulations and the further growth of squatter settlements (Kulaba, 1989). The author concludes that a significant legacy from the period (in addition to deterioration in services) was the 'stigma of lawlessness implanted into peoples' minds' by the abolition of urban councils, which proved difficult to erase even ten years after their reinstatement (Kulaba, 1989: 227).

The Local Government Act of 1982 provided the legal framework for the reinstatement of urban councils. Under the act, every local authority was to be governed by councillors elected from each ward within the town, members of parliament representing the town and five or six other members nominated by the Minister responsible for local government. The chief difference from before 1974 was that the mayor now had to be elected. Dar es Salaam City Council was re-constituted with 62 councillors, 52 of them elected from each of the 52 wards in the city. The four members of parliament for the city became co-opted members and the rest were nominated under special seats. The Council was presided over by an elected mayor assisted by a deputy mayor (Sheuya, 1996). However, the central government retained a firm hold. The director responsible for administration of council business and personally accountable for funds was appointed by the President. In addition, the central government had to approve all by-laws made by the councils, including proposals to change taxes. As the main source of finance for recurrent expenditure and the only source for the development budget, the central government retained tight control on spending.

Funds were generally inadequate, particularly the development budget. For example, the development budget for Dar es Salaam fell in real terms by 9% between 1978–1979 and 1986–1987 (Kulaba, 1989). The City Council was not able to raise sufficient funds to provide the services so desperately needed to meet the serious backlog. In 1989, Kulaba warned that urban local authorities would have to find new ways of raising revenue or they would face a financial crisis with urban services deteriorating even further.

The abolition of local councils was one of two mistakes Nyerere publicly admitted to in 1985 (Ngware, 1999). The other was the disbanding of cooperatives. Both were highlighted as useful instruments for popular participation. The return to the system of local government in 1982 had aimed to increase participation in development and local resource mobilization; however, neither goal was achieved (Therkildsen, 1998). The situation in Dar es Salaam was particularly bad. In 1996 the Dar es Salaam City Council was dissolved by the Government because it failed to discharge its duties. An appointed commission was put in its place. According to Nnkya (1999), this introduced a sense of purpose into the management of City Hall and within three years the Commission had made a difference in the collection of revenue, prompt payment of staff and improvement of services in the city (Nnyka, 1999). The Dar es Salaam City Commission had a limited life, however. One of its key tasks was to put institutional arrangements in place for the decentralization of the city into three municipalities. By 2000, four municipalities were established: the Dar es Salaam City Council (DCC) and the three municipal district councils of Ilala, Kinondoni and Temeke. The Dar es Salaam City Council is now composed of 20 councillors, who are drawn from the other municipal councils. DCC is a landless authority. Its role is coordination and operation of activities with a city wide scope.

These changes in Dar es Salaam were part of a broader movement to improve local government. In 1996, the Government set out its agenda for reform, followed by an action plan in 1997. The overall objective of the 'decentralization by devolution' policy was to improve the quality of, and access to, public services facilitated by local authorities. Government services expected to be delivered by local governments include: building and maintaining primary schools, hospitals, health centres, clinics and dispensaries, roads and water supplies. Urban authorities are also responsible for the establishment of fire brigades, public markets, slaughter houses, community centres, public parks, refuse collection and other local amenities[5]. The challenge was, and still is, to develop a local government system to deliver better services in a sustainable manner under conditions of extreme resource scarcity (Therkildsen, 1998).

In parallel with these changes, Government continued to implement policies aimed at transforming the centrally planned economy into a market-oriented economy. This entailed, in addition to privatization of public enterprises, a shift in the perception of the role of national and local government, from provider to facilitator of private and popular sectors in the development process (Nnyka, 1999).

5.3.2   Demographic and economic development in Dar es Salaam

Tanzania has experienced high rates of urban growth in recent decades. At the time of the 1988 census, 18% of the population on the Tanzanian

Table 5.8   Population of Tanzanian mainland, Zanzibar and Dar es Salaam

	1967	1978	1988	2002	78–88[1]	88–02[2]
Mainland	11,958,654	17,036,499	22,455,193	33,584,607	+2.8%	+2.9%
Zanzibar				981,754		
Dar es Salaam	356,86	843,090	1,360,850	2,497,940	+4.8%	+4.3%
Dar (% mainland pop.)	2.9%	4.9%	6%	7.4%		

Source: Census, various years: www.tanzania.co.tz/census.
[1] Average annual growth rate 1978–1988.
[2] Average annual growth rate, 1988–2002.

mainland was living in urban areas and this had grown to 24% (almost one in four) by the time of the latest census in 2002. The labour force surveys show that between 1990/91 and 2000/01, the Economically Active Population (EAP) in urban areas rose by 14.3% while rural areas saw a 3% decrease. Much of the urban growth in the past three to four decades has taken place in Dar es Salaam. In 1967 only 0.03% of the mainland population was living in the city, but by 1988 the figure had risen to 6%, and by 2002 to 7.5%. At the time of the 2000/01 survey, more than a million members of the EAP were in Dar es Salaam. The growth of the total mainland population and that of Dar es Salaam over the past 35 years is shown in Table 5.8. It can be seen that the population of the Dar es Salaam Region (94% of which is urban) was almost 2.5 million in 2002. If the average growth rate of 4.3% p.a. over the previous 12 years was maintained, the projected population in 2006 would be almost 3 million.

High rates of urban growth in Tanzania are associated with three phenomena:

- The growth of the informal sector
- Severe and chronic unemployment
- The proliferation of informal settlements

These phenomena and the responses to them by central and local government will be examined in the rest of this section.

5.3.3   The growth of the informal sector

Rapid growth in Dar es Salaam cannot be divorced from the growth of the informal sector. The distribution of the employed population by 'employment status' for Dar es Salaam and for the country as a whole is shown in Table 5.9. Clearly, agriculture has lost its predominant position in the city, with only 9% of the employed population working on their own farm. The largest proportion of the working population in Dar es Salaam is self-employed, with and without employees. They constitute almost half (48%)

Table 5.9   Distribution of employed population (standard definition) by employment status, 2000/01

	Paid employees	Self-employed and employees	Self-employed no employees	Unpaid family helper	Work on own farm	Total
Dar	291,547	40,989	306,788	33,130	66,266	738,731
Dar %	39%	6%	42%	4%	9%	100%
Country	1,150,055	159,685	1,241,151	644,763	13,639,376	16,835,031

Source: Integrated Labour Force Survey 2000/01 (URT, 2002).

of the workforce. It may be assumed that the majority of the self-employed are working in the informal sector. Paid employees are the next biggest group (39%) and some of these would also be working for informal operators. Thus, it appears that at least half of the working population of Dar es Salaam is employed in the informal sector, a minimum of 370,000 people.

This figure is remarkably close to the estimate of 345,869 that was produced by the Dar es Salaam Informal Sector Survey in 1995 (URT, 1998). The survey found that 222,915 (approximately two-thirds) of informal sector workers were operators and 122,954 were employees. Females made up less than half of the operators and only a quarter of the employees. Retail, eating and lodging was the most important sector by a long margin, both in terms of operators and employees. Manufacturing was second and construction third.

The Dar es Salaam Informal Sector Survey (URT, 1998) estimated the total number of construction workers in the informal sector in Dar es Salaam at 31,296[6]. This figure is roughly twice the number of construction workers employed in the formal sector in the city in 2000/01 (NBS, 2004). Roughly half of the informal workers were operators and half employees. There were 580 women estimated to be working in the sector, all of them as operators. It was found that 1200 operators had income from other activities, mostly wage employment in the private sector, but also a few in the public sector.

A number of factors have fuelled the growth of micro-enterprises and self-employment. For example, Tripp (1997) traces the development of informal income earning activities to the dramatic decline in real wages that began in the mid-1970s. Government was responsible for 65% of wages in the 1970s. Although wage increases were introduced periodically, high inflation meant that the effect of pay increases was quickly eroded. Real wages fell by 75% between 1974 and 1984 (URT, 1993), and continued to fall with the introduction of structural adjustment policies. In the late 1980s, it was estimated that the average worker's salary was sufficient to pay for three days of his/her household's monthly food budget (Tripp, 1997).

In this context, urban households had no option but to supplement their wages by engaging in other income earning activities; either in agriculture or in informal services. As wages fell, private earnings from work in the informal sector began to rise. The Household Budget Survey in 1977 found that wages accounted for 77% of total income in households where the head was a wage earner and private earnings were only 8%. By 1988, when Tripp (1997) undertook her survey, informal incomes were approximately 90% of the household's cash income with wage earnings only 10%.

Some households started to divide their time between wage-based and self-employment, often with women and children brought into the labour force to help out. Others decided to leave their jobs to concentrate full time on their private projects. According to the 1978 population census, 72% of workers in the city were in paid employment, but by 2000 the figure had fallen to just 39%. While the retrenchment of public sector workers (associated with structural adjustment policies) is generally held responsible for the loss of paid employment, Tripp (1997) shows that many in Dar es Salaam actually left their jobs voluntarily in the 1970s and 1980s in order to concentrate on their informal income earning activities. In Tripp's survey, 45% of the people left their jobs because of low pay and only 17% because of redundancy.

In sum, the phenomenal growth of the 'informal sector' in Tanzania (as in many other countries in sub-Saharan Africa) was largely a response to falling real wages as a result of general economic decline (Tripp, 1997). The vast masses of people working in the informal sector are individuals finding a way to make a few shillings in order to survive. Some of those working in the construction sector may have been retrenched from public service or laid off by private construction companies following the current trend of outsourcing labour. Some may still be in paid employment but wages are so low that they need to engage in other activities to supplement their incomes.

5.3.4   Government response to growth of the informal sector

There is some evidence that the difficulty of surviving in urban areas led to a decrease in rural to urban migration in the late 1980s. Life was also difficult in the rural areas, however, and many continued to find their way to the city. The 2004 Economic Survey estimates there are 650,000 to 750,000 new entrants to the labour force each year (URT, 2005a). Even without migration, more than 150,000 of these would be in the urban areas and 45,000 in Dar es Salaam. With migration, the annual increase is even higher. In this context, the informal sector and open unemployment have continued to grow.

The Government's reaction to growing urban unemployment and 'informal' activities was to bring in quite draconian measures to remove 'unproductive elements' from the towns and repatriate them to the rural areas – a practice inherited from the colonial regime.

The Human Resources Deployment Act, otherwise known as *Nguvu kazi*, was passed in 1983. This was a time when the economy was in serious trouble, resources of all kinds were hard to come by and destitution was widespread in the urban areas. The ostensible aim of the act was to ensure that every able-bodied adult be engaged in productive work (Kulaba, 1989). The act required all Tanzanians in urban areas to be registered and issued with labour identification cards. Those who could not produce proper identification were to be resettled in the countryside (Tripp, 1997). The police, army and people's militia in Dar es Salaam immediately started rounding up all those suspected of 'loitering', which included petty traders and all those who did not have a job or a licence. Those declared unemployed were sent to their home villages or to government sisal estates.

However, it was clear from the start that the campaign could not possibly succeed, not least because it antagonized the majority of the population, who had no choice but to try to make a living through their own informal projects. By 1984, the City Council changed its position and began to register people engaged in informal activities through its ten-house cell system. It also approved and issued licences to some income-generating activities conducted on locations approved by the Council. The Government also began to recognize the importance of small-scale enterprises to the national economy. By 1994, it had developed a national policy to assist informal activities in both urban and rural areas, and this began to attract donor support. The Human Resources Deployment Act came to be seen as a turning point in public policy because the issue of income-generating activities outside the formal economy was faced for the first time.

Once the right of small business to exist was accepted, the debate shifted to the issue of how municipal authorities dealt with licensing and businesses location problems. The 1982 Urban Authorities Act gave local authorities the power to determine how operators should function. After 1985, 56 different kinds of small business activities required licences to operate in Dar es Salaam (Tripp, 1997). Although micro-enterprises considered legalization of their activities an improvement on the previous policy, regulation through licensing brought new problems. Some were charged and many more detained for violating licensing regulations. They were rounded up by militia hired by the City Council. Up to the mid-1990s, the Dar es Salaam City Council militias continued to harass and round up street vendors in order to move them to different locations. The vendors maintained that these City Council employees were seizing their goods for personal gain and extracting bribes from those without licences. The sympathy of the public was with the vendors, who also received support from the central government and the lower levels of the party hierarchy (Tripp, 1997). These actions may have been one factor in the Government's decision to disband the Council in 1996 and appoint a commission in its place.

From the mid-1990s, Government produced a number of policies which clearly recognized the role of the informal sector. However, it was still encouraging people to move back to the countryside in order to ease the problems of urban growth. The first aim of the National Employment Policy, approved by the cabinet in 1997, was: 'to prepare a conducive environment for the unemployed to employ themselves by directing more resources to the self-employment sectors' (URT, 1997: 4). Additional aims were: 'to develop the self-employment sector in rural areas so as to reduce the rate of migration to urban areas' and 'to direct most of the labour force into the currently unattractive sectors...such sectors are agriculture, livestock and cooperatives' (URT, 1997: 4–6). This objective was to be achieved by the decentralization of industrial investment away from the major towns (particularly Dar es Salaam) and by advising young men in urban areas to move to rural areas where they could use their skills and employ themselves. Local authorities were to be at the forefront in implementing this employment policy. They were to create departments of employment and youth development, and appoint labour and youth officers to 'sensitize people to initiate self-employment' (URT, 1997: 417). The way in which the city of Dar es Salaam is implementing this policy will be described in Section 5.4.

5.3.5   Growth of informal settlements and government response

A second major manifestation of rapid urban growth, combined with limited resources, has been greatly increased pressure on urban infrastructure and services. Urban growth has not been matched in the availability of houses or plots in planned areas and unplanned settlements have grown steadily over the past 30 to 40 years in Dar es Salaam and other major towns in Tanzania.

A survey of 660 houses in Dar es Salaam in the late 1980s revealed that 89% relied on simple pit latrines for sanitation and 60% had to share with other families. Pit latrines present a serious problem during the rainy season when the water table rises, particularly in low-lying areas and river valleys, commonly occupied by squatters. In the households surveyed by Kulaba (1989) 21% reported overflowing latrines. Refuse collection was another major problem. Government data from 1986 estimated that only about one-fifth of refuse generated in Dar es Salaam was collected. Urban roads were also in an appalling state due more to shortages of equipment than manpower. The budget for road maintenance and rehabilitation in 1986–1987 was estimated at 0.4% of the sum required.

The situation has hardly improved. The majority of city residents still live in unplanned settlements which lack basic infrastructure (roads, surface water drainage, water supply, sanitation and waste disposal). According to the 2000 *Household Budget Survey*, 82% of the population of Dar es Salaam were using pit latrines and a further 6% had no toilet at all (URT, 2003). Even the planned residential areas are not all fully serviced.

As elsewhere in the developing world, unplanned settlements in Dar es Salaam tend to be regarded as slums and the initial Government response was to demolish them. The first slum clearance program in Tanzania was carried out in Buguruni (Sheuya, 1996). But by 1970 it was realized that the approach was too costly and was replaced by a policy of upgrading unplanned settlements and providing new serviced plots. Upgrading projects were implemented between 1974 and 1981 in Dar es Salaam and six other towns, with funds from the World Bank. The projects were implemented by the Sites and Services Project Unit under the Ministry of Lands, Housing and Urban Development in the central government. The approach was incorporated into Tanzania's housing policy in 1981, which recognized that the Government could do little to help in the housing field and people simply had to make their own provision (otherwise known as an enabling approach). However, the policy said little about the provision of services, which remained the responsibility of the Ministry of Lands. The involvement of this ministry in the provision of serviced plots in the city is ongoing.

It was not until 1992 that Tanzania adopted a national strategy for human settlement development with the twin objective of improving living conditions in informal settlements and alleviating poverty through the stimulation of economic growth and employment. A national program to implement this strategy was subsequently put in place with assistance from international donors. In line with current thinking, the strategy focused on strengthening the capacities of local authorities, NGOs and the private sector. Upgrading plans were to be prepared and implemented by local authorities with the participation of residents and local communities. Local resources were also to be mobilized to finance these plans through appropriate cost recovery (Sheuya, 1996). The second phase of the Sustainable Dar es Salaam Project and its extension to other urban centres was part of this program, as will be explained in Section 5.5.

5.4   Decent work indicators

5.4.1   *Indicators of employment*

Unemployment

According to the internationally accepted standard definition, a person is classed as unemployed if s/he is (a) without work, (b) available for work and (c) looking for work. This is called category A unemployment. However, the standard definition allows relaxation of condition (c) in countries where a large proportion of the population is in subsistence agriculture or informal activities with little knowledge of labour market developments in the rest of the country. Hence, according to the *Relaxed Standard Definition* (called the *Standard Definition* in the Tanzanian Labour Force survey), a person is unemployed if s/he fulfils criteria (a) and (b). This is known as Category B unemployment.

Summary of data on employment

According to the standard definition of unemployment:

- In 1990/91 3.6% of the economically active population in mainland Tanzania were unemployed, increasing to 5.1% in 2000/01.
- In the urban areas, unemployment was 10.6% in 1990/91, increasing to 14.8% in 2000–20/01.
- Unemployment in Dar es Salaam was 26.4% in 2000/01 by the standard definition, but 46.5% by the Tanzanian definition: there is no data for unemployment in Dar es Salaam for 1990.
- There is no data for unemployment in construction, but it is estimated that 50% of those who are self-employed in the construction industry are unemployed for the majority of the time.

The Tanzanian *National Definition* (which is country specific) includes a third group of unemployed, those without sustainable employment or with only a marginal attachment to work. In the Tanzanian context, it is often the case that work has been done during the reference period but it is not work that is accessible to that person from day to day. When this is the case 'it is not reasonable to consider it as employment, as there is a possibility that the person is actually unemployed for most of the time' (URT, 2002: 72). This category of unemployment, known as category C, is applied only to those who are self-employed and family helpers in non-agricultural activities, Inclusion in this category is at the discretion of the enumerators.

A breakdown of the unemployed in mainland Tanzania by category and geographical area is shown in Table 5.10. Unemployment rates based on the National Definition, which includes all categories, are shown in Table 5.11.

Table 5.10   Number of unemployed persons by category of unemployment and location, 2000/01

Category	Dar es Salaam	Other urban	Rural	Total mainland
(A) Looking for work	207,158	107,954	97,491	412,603
(B) Available not looking for work	57,651	134,043	308,475	500,169
(C) Marginally attached to work	202,290	384,344	806,942	1,393,576
All categories	467,099	626,341	1,212,908	2,306,348

Source: *Integrated Labour Force Survey 2000/01* (URT, 2002).

Table 5.11   Unemployment rate (percentage) by category of unemployment and location, 2000/01

Category	Dar es Salaam	Other urban	All urban	Rural	Total mainland
(A) Looking for work	20.6	4.4		0.7	2.3
(B) Available, not looking for work	5.7	5.5		2.1	2.8
(A + B) by standard definition	26.4	9.9	14.8	2.8	5.1
Comparison for 1990/91 A+B			*10.6*		*3.6*
(C) Marginally attached to work	20.1	15.9		5.6	7.8
(A + B + C) by Tanzanian definition	46.5	25.9		8.4	12.9%

Sources: Labour Force Survey 1990/91; Integrated Labour Force Survey 2000/01 (URT, 1993; 2002).

Table 5.11 indicates that the overall rate of unemployment, even according to the national definition, is quite moderate at 12.9%, but there are very wide variations between locations. Rural areas have the lowest rate at 8.4%. In urban areas, excluding Dar es Salaam, the rate is three times higher at 25.9%. The unemployment rate is highest in Dar es Salaam at a staggering 46.5%.

The standard definition of unemployment is included in the 2000/01 survey so as to enable comparison with 1990/91. By this definition there was an increase in unemployment in the ten-year period. In 1990/91, 3.6% of the economically active population were unemployed by the standard definition (A + B), compared with 5.1% in 2000/01. In urban areas, unemployment rose from 10.6% in 1990/91 to 14.8% in 2000/01. In the ten-year period, the absolute number of unemployed in urban areas increased by 152%, and in rural areas by 98%.

There is no data on unemployment in the construction sector. However, it is interesting to note that the number of self-employed construction workers is 66,943 according to the *standard definition* of employment. According to the *national definition*, however, the number of self-employed in the construction workforce was only 34,182. This suggests that around half of those who are self-employed in the construction industry are only marginally attached to their employment and have been classed as unemployed (category C) according to the Tanzanian definition.

Low wages

Minimum wages are set by a minimum wage board comprising the social partner representatives (employers and workers) under the auspices of the Ministry of Labour. This is provided for under the 1953 Regulation of Wages and Terms of Employment Ordinance. Once the applicable minimum wage is proclaimed, by way of minimum wage orders, it becomes binding. Any worker, or even labour officer, can in principle prosecute defaulting employers in court. Subject to contracts of service, wages are supposed to be paid in full and on time.

The minimum wages from 28 June 2002 until they were revised in 2007 were as follows:

Hourly	Tsh250
Daily	Tsh1846
Monthly	Tsh48,000[7]

Summary of data on low wages

In mainland Tanzania in 2000–2001:

- 26% of paid employees received less than half the median wage, compared with 8% in 1990/91.
- 57% of the self-employed are estimated to have received less than half the median wage, compared with an estimate of 56% in 1990/91.
- There is no data for construction, but 65% of those in 'elementary occupations' (which includes construction workers) in 2000/01 earned less than half the median wage.

In addition:

- Many of the 75% of construction workers employed as 'casuals' in the private sector in 2000/01 received less than the minimum wage.
- The average wage for regular construction workers was more than double that of casuals in 2000/01.

The minimum wage is not updated very often and information on the level set in earlier years and the dates at which it changed was extremely hard to find. Sheuya (1996) gives the minimum wage in 1996 as Tsh30,500 and before that it was Tsh17,500. The minimum wage orders which are proclaimed under the above-mentioned ordinance apply to both the private sector and the employment of casual workers by national or local government. They do not apply to rates payable by the Government to its civil servants, which are customarily proclaimed by Government circulars and not meant for public consumption. Thus, it is common to observe a difference (albeit not great) in the minimum wages payable to workers in the construction industry who are civil servants, as compared to those working in the private sector. However, there are currently very few of the former (ILO, 2005b).

The minimum wage rates currently apply to all sectors, including construction, and are uniform throughout the country. However, under new labour legislation (the Labour Institutions Act of 2004) the Minister may appoint sectoral wages boards to investigate wages and terms of employment, and make recommendations on minimum wages for a particular sector or area.

Table 5.12   Median income of paid employees among the employed population by main sector, 2000/01

Government	Parastatal	Traditional agriculture	Private informal	Other private
65,000	95,000	7500	10,000	30,000

Source: *Integrated Labour Force Survey 2000/01* (URT, 2002).

Low wage indicators

According to the ILFS 2000/01, the median monthly income for paid employees in all industries was Tsh30,000 per month. It was not possible to ascertain from the survey data what proportion of the employed population was earning less than half the median wage, which would be less than Tsh15,000. But we can calculate from the available data that 23% of paid employees were receiving less than Tsh10,000 and 29% were receiving less than Tsh20,000. On this basis it may be assumed that roughly 26% were receiving less than Tsh15,000. The comparable figure for 1990/91 was 8%.

The median income of the self-employed population in 2000/01 was Tsh17,143. Distribution around the median incomes is not shown, but distribution around the mean shows 57% of self-employed workers with a monthly income of less than Tsh10,000. The figure is similar (56%) for 1990/91.

The median income of those in paid employment in construction in both 2000/01 and 1990/91 was somewhat higher than the median for all industries, while the median income for the self-employed in construction was double the median for all industries. We do not know what proportion of paid employees were earning less than half the median income. We can, however, calculate from available data that 65% of workers in the category 'elementary occupations' in 2000/01 were earning less than half the median. Many of these workers may be working in the construction industry. There is no information on the distribution of income from self-employment in construction.

The median income by main sector according to the 2000/01 ILFS is shown in Table 5.12. It can be seen that Government employees, particularly parastatal employees, earn more than those working in the private sector and much more than those working in the informal sector. The Employment and Earnings Survey, 2001 (NBS, 2004) had a similar finding. The overall average monthly wage of those in regular employment in the public sector was Tsh123,779, while in the private sector it was Tsh108,266. Central and local government employees accounted for slightly over one-third of total employment but earned 49% of total earnings (NBS, 2004: 32).

According to the same source, the median wage of those employed in formal establishments in construction was almost the same in the public and private sectors at Tsh55,352 and Tsh50,853 respectively. The

Table 5.13   Hours worked and payment received on 11 large construction projects

Project number	Normal working hours/week	Normal working hours per month	Required salary for four weeks work	Actual salary	Under payment
1	60	240	60,000	40,000	20,000
	72	*288*	*72,000*	*36,000*	*36,000*
2	48	192	48,000	45,000	3,000
	65	*260*	*65,000*		*20,000*
3	60	240	60,000	45,000	15,000
4	60	240	60,000	40,000 +	20,00
	84	*336*	*84,000*	*36,000*	*48,000*
5	51	204	51,000	50,000 +	Nil
6	56	224	56,000	36,000	20,000
	84	*336*	*84,000*		*48,000*
7	54	216	54,000	60,000 +	Nil
8	74	296	74,000	54,000	20,000
9	48	192	48,000	n/a	
10	42.5	170	42,500	45,000	Nil
				60,000	Nil
11	48	192	48,000	48,000	
	72	*288*	*72,000*	*36,000*	*36,000*

Source: Baseline Study of Labour Practices on Large Construction Sites in Tanzania (ILO, 2005b). The figures in italics represent the workers' estimates where these were available: the higher salaries claimed by workers on project number ten are attributed to overtime payment, which the workers agree is paid on this project.

construction industry paid the lowest wage in the public sector, but in the private sector, wages of formal sector workers in agriculture and commerce were lower than in construction. However, these averages mask a big discrepancy between the wages of regular and casual workers. The average wage of regular construction workers at Tsh92,342 was more than double the average wage of casual construction workers at Tsh39,173 (NBS, 2004). In 2001, average casual earnings ranged from Tsh18,694 in non-profit private firms to Tsh35,148 in Government and Tsh39,509 in profit-making parastatals.

All of the contractors interviewed in the baseline study of labour conditions on 11 large construction sites that was commissioned by the ILO in 2003 (ILO, 2005b) claimed to adhere to Government regulations on pay. However, the two contracts for which information on wages was provided by both employers and workers showed a large gap between what the employers claimed to pay and what the workers said they actually received. In one instance workers said they received only 50% of what the contractors claimed to be paying. On the basis of the workers' evidence, casual workers on three road projects (1, 4 and 6 in Table 5.13) were receiving only Tsh36,000 per month, which is clearly below the minimum wage of Tsh45,000. These three projects also had the longest working hours and on two of the three no overtime was paid.

On the eight other projects it was not possible to confirm that the minimum wage of Tsh45,000 was actually paid. However, even if it was, the minimum monthly wage is calculated on the basis of the hourly rate and assumes a normal working week of 45 hours (nine hours per day for five days) and a normal month of 180 hours, whereas on most of the projects, the normal working week was far in excess of 45 hours. In these cases the minimum legal wage requirement was not being met. Taking into account the number of hours worked in a 'normal' week, it would appear that on only three projects were casual labourers being paid the legal minimum wage or above. On the majority of projects, contractors were failing, on their own evidence, to comply with the law on minimum wages and workers were being seriously underpaid. The information on which this conclusion is based is shown in Table 5.13.

The situation is very different for casual workers employed directly by the Dar es Salaam City Council. They are paid at the official minimum rate of Tsh1850 per day and work for only eight hours. There are very few such workers, however. At the time of the interview there were 16 employed by the Council on regular maintenance work at Karimjee Hall, about 12 at the city council dump site and a few at the parking lot. The number of directly employed manual workers had declined by about 30–40% in the city as a whole.

Hours of work

The Regulation of Wages and Terms of Employment Ordinance (1953) states that normal working hours are 45 per week (nine hours per day for five working days). Should an employee work for more than the above prescribed hours, additional wages are to be paid in compensation. The current rate of overtime pay is 1.5 times the hourly wage for every hour worked on weekdays and twice the hourly wage for every hour worked on Sundays and public holidays.

Summary of data on hours of work

In 1990/91 in mainland Tanzania:

- 42% of paid employees and 54% of the self-employed worked more than 49 hours per week.
- Construction workers reported usual working hours of 51 per week.

In 2000/01:

- 65% of paid employees and 61% of the self-employed worked more than 49 hours.
- Construction workers reported *usual* working hours of 62 per week.
- The ILO survey in 2003 found hours worked in excess of 70 on one-third of major projects.
- The Informal Sector Survey in Dar es Salaam in 1995 reported average working hours in construction of 56 per week.

Data on compliance

The proportion of wage earners working for more than 48 hours per week could not be obtained. However, the labour force survey of 2000/01 found that 26% of the total employed population was working for more than 49 hours, compared with 1.9% in 1990/91. Long hours of work predominate amongst those working outside agriculture. It was found that 65% of paid employees and 61% of the self-employed worked for more than 49 hours in 2000/01. The comparative figures for 1990/91 were 42% and 54% respectively.

Transport was the sector with the highest number of hours worked in 2000/01 and construction was second. Construction workers reported usual working hours of 62 per week, compared with 36 and 47 for all industries. This was a marked change from 1990/91, when the usual number of hours worked in construction was 51 and this sector was joint fourth after transport, trade and mining.

The 2003 survey of 11 large construction sites found that working hours on seven of the eleven projects were well in excess of 45 per week. The maximum number of hours worked per week was 84 on two projects and 72 hours on two other projects. Moreover, on two of these four projects no overtime was paid. Casual workers do not generally receive days off with pay: paid leave is for permanent workers only. Hours worked in the informal sector are also long. The Dar es Salaam Informal Sector Survey of 1995 found that temporary and casual employees in construction had worked on average for 232 and 206 hours respectively in the previous month, meaning 50 to 60 hours per week. These long hours can indicate low wages, that is, workers having to do long hours in order to make an acceptable income. It also indicates low productivity.

Occupational health and safety

> **Summary of data on OHS**
>
> - There is no data on working days lost due to accidents in the workplace in Tanzania.
> - A special survey in 2001 revealed 150 accidents in the previous 12 months on 63 large building sites in mainland Tanzania.
> - An ILO study of major projects in mainland Tanzania in 2003 revealed non-compliance with H&S legislation on 64% of projects.
> - In Dar es Salaam the 1995 informal sector survey revealed: five accidents per hundred workers in all sectors and thirteen accidents per hundred workers in construction.

The law is quite explicit on occupational health and safety (H&S). The 1999 Employment Ordinance Act No. 9 dedicates the whole of part VIII to set out provisions on health and safety. More specifically, section 99 requires an adequate supply of water for all employees and members of their families living with them on the employer's property, while section 100 requires that they be provided with medical care (ILO, 2005b).

The construction industry has particular rules on health and safety, contained in the Factories (Building Operations and Works of Engineering Construction) Rules, which are mandatory in the construction industry. They require that contractors cooperate to ensure safe working conditions, prepare a written health and safety policy statement including arrangements for carrying it out and for bringing it to the notice of employees, appoint a safety officer on sites with more than 20 workers and notify the chief inspector of the name of the safety officer. The contractor is also required to provide fresh drinking water, suitable accommodation for preparing meals and drying clothes, appropriate facilities for washing and sanitation, and first aid. There are technical requirements for scaffolds, guardrails, working platforms, etc. There is a requirement that employers take out compulsory insurance against injuries and fatal accidents that may occur at worksites.

Additional contractors' obligations are set in the Contractors Registration Board (CRB) by-laws (1999) under the Contractors Registration Act of 1997, as follows.

Contractors are obliged:

- To provide and maintain appropriate safety gear (personal protective equipment) for all persons on site
- To maintain a register with a record of all accidents and injuries
- To display a signboard with names of the client, consultants and contractors and a hoarding (to prevent accidents to passers by)
- To provide firefighting equipment and hygienic facilities

In 2003, Tanzania passed a new Occupational Health and Safety Act which provides for the establishment of health and safety representatives, as well as their training and duties to be carried out in working hours. It also requires health and safety committees at factories and other workplaces with more than 50 workers. The Act provides for the establishment of an Occupational Safety and Health Authority (OSHA) based on the old factory inspectorate. At the time of writing, the OSHA has been established but the act itself is under revision and guidelines for its implementation have still to be drawn up. It is also intended to update regulations for health and safety on construction sites, possibly along the lines of the South African regulations, which place ultimate responsibility upon the client, and share the burden between stakeholders, as opposed to placing all responsibility on the contractor.

Data on compliance

There is no data on working days lost due to accidents in the workplace in Tanzania, at national or city level, and no data on accidents or deaths in the construction industry. However, even casual surveillance of construction sites confirms that the standard is very low and the requirements are not being met. A survey of health and safety in the construction industry was undertaken for the Contractors Registration Board (CRB) by the Occupational Safety and Health Authority (OSHA) in 2001 (CRB, 2001). A total of 150 accidents in the previous 12 months were recorded by the survey team on the 63 sites visited, three of which were fatal.

Most of the contractors interviewed in the study of working conditions on 11 major sites in 2003 claimed to provide all of the entitlements and take all of the precautions required by law (ILO, 2005b). These include the provision of protective gear, first aid facilities, transport to the nearest medical centre in case of emergency, training on H&S, presence of a safety officer, insurance provision, recording of accidents and compensation to injured workers. However, interviews with workers revealed that these regulations are complied with on only a few of the sites. Only five of the eleven sites visited had a safety officer, only three sites provided any training to the workers and only two sites had H&S representatives and/or committees. The record for protective gear (the easiest thing to comply with) was just as bad. On only four of the eleven sites was the contractor providing all of the necessary protective gear. On six sites helmets were provided, but only to a few employees. On one site no protective gear of any kind was in evidence.

Most sites did, however, have a first aid box and about half had provided some first aid training, although this was generally felt to be inadequate. A few contractors had paid for treatment and offered compensation to workers who had suffered injuries at work, although workers are generally unaware of their entitlement in this regard. Many of those interviewed expressed a desire for more information on their entitlements and for training on H&S and HIV/AIDS. The research concluded that the 11 projects exhibited a very wide range in the standard of occupational safety and health (OSH). Overall, an acceptable standard was achieved on four of the projects. On four other projects the provision for OSH was inadequate, and on three further projects it was exceptionally poor. Two of the three worst projects were under Chinese contractors. The National Housing Corporation was acting as both client and contractor on the third.

The situation was equally unsatisfactory in terms of workers' welfare. Most of the contractors claimed to provide accommodation, transport, food and water to all of their workers, but evidence provided by the workers showed that this was seldom the case. On only two out of seven projects in rural areas were fresh drinking water and toilets provided. In most cases the workers reported that they had to use the surrounding bushes for toilets and ask local villagers for water. Only one project provided washing and changing facilities. Accommodation was only provided on two rural projects, and only for permanent or skilled workers.

The most common health hazards reported by operators in the Dar es Salaam informal sector survey were dust, noise, poor water supply, extreme temperatures, poor waste disposal and inadequate working space. Construction had an above average rating for health problems and was among the four sectors with the highest accident rates. During the previous 12 months 2023 accidents were reported in among 15,009 construction workers, giving a rate of 13 accidents per 100 workers. The comparative figure for all sectors combined was five (URT, 1998: 1–82).

The UNV/ILO project with informal construction workers in Dar es Salaam reinforced these findings (Jason, 2005). Workers are aware of a number of health problems related to their work. For example, carpenters and stone crushers think they are at high risk of lung diseases due to exposure to dust. They are also concerned about safety. Informal construction workers on the project are, however, generally hired directly by private house-owners who cannot be held responsible for the health and safety of the workers and cannot be expected to provide them with protective gear. Sometimes they also work for contractors, but health and safety on construction sites in most cases is limited to advising caution, as opposed to attempting to eliminate hazards or providing workers with protective gear (Jason, 2005).

5.4.2 Indicators of social security

Public social security coverage

The National Social Security Fund Act, No. 28 of 1997 established the National Social Security Fund (NSSF), which took over the assets, liabilities and customers of its predecessor, the National Provident Fund. The NSSF Act refers to three groups of persons who it will insure:

(1) Every person who is a member of the existing fund
(2) Every person who is self-employed or employed in the private sector other than in a body which is a parastatal organization under the Parastatal Pension Fund Act, 1978
(3) Every non-pensionable employee in the Government service and parastatal organization (URT, 2002: 8)

Any other group may be included – or excluded – under orders of the Minister.

Perhaps more interesting than those included are those left out. Apart from the public sector employees, who have their own fund (see below) all *temporary employees* are excluded. Since the majority of workers in the construction industry (as well as many other sectors) are employed as casuals, often for as little as a day, they are automatically excluded from membership of the NSSF. Strictly speaking, the law allows employers to hire workers on a casual basis, although social security legislation and the Employment Ordinance both discourage casualization of employment relations. Under the Employment Ordinance, a casual employee is defined as one who is paid at the end of each day. Any such employee who works for 280 days in a given year is entitled to all the benefits enjoyed by other contractual employees. However, in practice this is seldom the case.

The NSSF is a contributory scheme, with employers paying 20% of wages and employees paying 10%. Employers are empowered to deduct the employees' contribution from wages. It is not entirely clear how much self-employed persons should pay, but it seems that they must have special permission and the sum is negotiable; but not less than the employees' contribution and not more than that normally contributed by the employer. The fund provides benefits in the following classes: retirement pension, invalidity pension, survivor's pension, funeral grants, maternity benefit, employment injury benefit and health insurance benefit. There is no provision for unemployment.

Summary of data on social security coverage

- There is no provision for unemployment benefit or sick leave in Tanzania.
- Formal social security schemes provide only for old age, maternity and invalidity benefits.

Old age pension

Summary of data on pension coverage

- No data is available on the proportion of the population over 65 without pension coverage.
- Only 3% of the total employed population is estimated to have been covered by pension schemes in 2003.
- The number of construction workers covered is not known.

According to the Economic Survey of 2004, there were 327,609 workers enrolled in the NSSF in 2002–2003. This is 43% of those working in 'other' private activities in 2000/01, 15% of those working in the private sector (including the informal sector) and only 1.9% of all employed persons. The total membership at the beginning of 2006 was 400,000.

Coverage is greater in the public sector, where the Public Service Retirement Act of 1999 replaced the pension ordinance of 1954. Members of the Parastatal Pension Fund (PPF) are pensionable employees of the Government and its executive agencies. The Economic Survey 2004 reported 196,307 members of the Fund in 2003, increasing to 205,681 members in 2003–2004. The first figure represents 46% of the total number of employees in the public sector (government and parastatal) in 2000/01. Taking both public and private sectors together, a total of 523,916 persons were estimated to be covered by these social security schemes in 2003. This is 3% of the total employed population, or 20% of the total workforce outside the traditional agriculture and household sectors in 2000/01 (as shown in Table 5.4).

Unfortunately, the available data is not broken down by industry. The majority of contractors interviewed in the 2003 survey of 11 construction sites claimed that their workers were registered with the NSSF. Many claimed that all workers were registered but others that only some were. This was usually the permanent workers. In almost all cases the labourers who form the majority of the workforce were not covered.

Workers and employers were found to be ignorant of the reason for construction workers' exclusion from the NSSF. Various reasons were put forward, including that their monthly salary was too low. On one site it was suggested that to be registered with the NSSF, you had to have a salary or more than Tsh50,000 per month, and no Tanzanians on that site were being paid that much so none registered. On another site, failure to register with the NSSF was attributed to the fact that workers are paid by task. Workers also thought they might be excluded from the scheme because they did not have a written contract. None of the construction workers covered in the Dar es Salaam Informal Sector Survey had a written contract.

Workers interviewed on one site, who were registered with the NSSF, were uncertain of their benefits. Newspaper reports on misuse of funds by the NSSF reinforce these uncertainties[8]. In addition, employers' capacity to make deductions on behalf of workers while failing to register them was mentioned in interviews with workers. A similar complaint of failing to forward contributions has even been made against the construction workers union (TAMICO) by its employees.

5.4.3 Indicators of workers' rights

Under the Constitution of the United Republic of Tanzania, discriminatory treatment in terms of wages and related privileges at workplaces is expressly prohibited. A victim of any discriminatory practice can obtain remedy in courts of law, either the Industrial Court of Tanzania or the High Court. Apart from the Constitution, there is no other legal provision on discriminatory practices currently in force. However, the new Employment and Labour Relations Act, 2003 has a provision that outlaws discriminatory treatment in workplaces in the following terms: 'S. 7 (1) every employer shall ensure that he promotes an equal opportunity in employment and strives to eliminate discrimination in any employment policy and practice.'

Wage inequalities between genders

> **Summary of data on discrimination**
>
> - In mainland Tanzania the median monthly income of paid male employees was 20% higher than that of females in both 1990/01 and 2000/01.
> - The median income of self-employed males was 164% higher than that of self-employed females in 1990/91 and 80% higher in 2000/01.
>
> In the construction sector:
>
> - The median income of paid male employees was 25% higher than that of females in 1990/91.
> - The median income of paid female employees was 43% higher than that of paid males in 2000/01.
> - There were no reported self-employed females in construction in 1990/91.
> - The median income of self-employed females in 2000/01 was 20% higher than that of males.
> - There is no data on wage inequality according to place of birth.

There is some discrepancy between the wages paid to men and women in Tanzania. According to Labour Force Surveys, the median monthly income for male paid employees was 20% higher than for females in both 1990/91 and 2000/01. The gap was much wider among the self-employed. In 1990/91 the median income of self-employed males was 164% higher than that of self-employed females. The gap had narrowed by 2000/01, however, when self-employed males earned 80% more than females.

The Employment and Earnings Survey of 2001, which covered only formal sector employees, also reported a gap between the pay of male and female workers in both the Government and private sectors. Parastatal profit-making was the only sector where male and female earnings were almost equal.

In 1990/91, the median income of male paid employees in the construction industry was 25% higher than that of females. However, in the Labour Force Survey of 2000/01 the positions are reversed. Female employees earned more than male employees by a considerable margin. The median earnings of female paid employees were Tsh50,000 per month, whilst for males only Tsh35,000. The most probable explanation lies in the fact that women and men are doing different jobs in the construction industry. The construction workforce in Tanzania is 97–98% male. Women are employed in administrative and secretarial functions only, occupations requiring some education and earning a higher salary.

A similar pattern can be found amongst the self-employed working in construction in 2000/01, with women earning Tsh42,857 and men Tsh35,000, which is not such a large gap but still significant. This is more difficult to explain. The most probable explanation is, once again, that women were performing different tasks from men – in this instance, probably preparing and serving food on the construction sites. The fact that the median earnings for male and female workers combined is the same as for males indicates that the proportion of self-employed women working in construction is very small. In 1990 there were none recorded.

Wage inequalities according to place of birth

There is no data on wages by place of birth in Tanzania. However, the ILO survey of 11 construction sites found a very great difference in the terms of employment between Tanzanians and foreigners (ILO, 2005b). Most of the foreign workers (along with a few Tanzanians) were on permanent contracts, while all of the casual and temporary labourers were Tanzanians. Not only wages but also benefits such as holidays with pay, insurance, pensions, housing, etc. are very different between these two forms of contract.

Child labour

The term 'child' is defined under section 77 of the Employment Ordinance (1957) as a person under 12 years of age. The ordinance forbids employment of a child under the age of 12 in any capacity whatsoever. Anyone employing a child under this age is guilty of an offence. However, as the law stands at present, a person over 12 years old may be employed in certain categories of work. If the employee is less than 18 s/he must return home to his/her parents or guardian every day after work and should not be employed near machinery or in work considered injurious to the health.

Summary of data on child labour

- Twenty-three per cent of children aged 10–14 were economically active in mainland Tanzania in 1990/91.
- Fifty-three per cent of children aged 10–17 were economically active in 2000/01.
- In Dar es Salaam, 14% of those aged 10–17 were economically active in 2000/01 and most were in the 14–17 age bracket.
- The construction industry employed 407 children aged 10–14 in 2000.
- The informal sector survey (1995) found no children working in construction in Dar es Salaam.

The Employment and Labour Relations Act, 2003, makes it illegal to employ a child under 14 years of age. In general terms, the Act suggests that children should not be employed in activities that are inappropriate for their age, that is, in activities that place the child's well-being, education, health or social development at risk. For example, a child of 14 can be employed only in light work. A child under the age of 18 cannot be employed in a mine, factory or as crew on a ship, unless it is part of his/her training.

Data on compliance

The ILFS 2000/01 found that 53% of children aged ten to seventeen are economically active in the country as a whole, a total of 3,463,716. It is not known how many were under the current legal age of 12 or the legal age of 14 under the new act. However, in 1990/91, 23% of 10 to 14-year-olds were economically active, compared with 65% of 15 to 19-year-olds. Hence, it may be assumed that a high proportion of the children working in 2000/01 were also in the older age group. The highest proportion of working children (68%) and by far the greatest number are in rural areas. In urban areas, 28% of children aged ten to seventeen are economically active. In Dar es Salaam the figure is 16%.

The 2000/01 survey found that the construction industry employs 1977 of the total number of working children, all of them male. These include 1618 paid employees and 358 unpaid family helpers. A special report on child labour in Tanzania found that most (1462) of the children working in construction were aged 15–17, which is legal in Tanzania. However, 407 were aged ten to fourteen and 105 were under nine (URT, 2001). The Dar es Salaam Informal Sector Survey reported no children working in construction in the city.

5.4.4   Indicators of social dialogue

Institutional structure

Labour relations in Tanzania are regulated by legislation and supervised by the Ministry of Labour and Youth Development in partnership with the trade unions and employers' organizations. The trade unions derive their legitimacy under the provisions of the Trade Unions Act, 1998. The employers are represented by the Association of Tanzania Employers (ATE), which is registered as an association under the same act.

Under the Ministry of Labour three key departments can be distinguished, namely:

- The Office of the Commissioner for Labour, which has overall responsibility for supervising labour relations in the country
- The Office of the Chief Inspector of Factories and Other Workplaces, which has recently been reconstituted as the Occupational Safety

and Health Authority appointed under section 4 of the Occupational Health and Safety Act, 2003
- The National Employment Promotion Service, established under section 3 of the National Employment Promotion Service Act

The Office of the Commissioner for Labour and Labour Officers working under the Commissioner constitute the institutional structure for the resolution of workplace grievances before they are referred for adjudication or reconciliation. The law provides for adjudication through the Industrial Court of Tanzania (established under the Industrial Court of Tanzania Act, 1967). Ordinary civil courts are also empowered to deal with labour related disputes or grievances, albeit limited to reports submitted in court by labour officers. Conciliation of workplace grievances or disputes at workplaces may also be handled by conciliation boards, normally established at district levels.

Under the new labour laws (the Employment and Labour Relations Act, No. 6 of 2004 and the Labour Institutions Act, No. 7 of 2004) a Commission for Mediation and Arbitration will be set up to intervene on labour disputes and grievances. The Industrial Court will be replaced by a Labour Court which will in turn be a division of the High Court.

Trade union density rates

The law is permissive regarding the formation of trade unions. A minimum of 20 employee signatures are needed to form a trade union. In principle, multiple unions can be formed for any particular trade or industry. However, all unions must be registered and the registrar is a Government appointee with significant power, including the power to refuse registration. Where two or more trade unions exist in an establishment, occupation, trade or industry, the registrar can cancel the registration of all except that with the largest number of employees, if s/he considers it to be in their interests. S/he can also require the unions, other than the one with the largest number of members, to cancel the membership of workers in that establishment, trade, occupation or industry. Hence, the registrar has the power, in practice, to ensure that there is only one trade union representing workers.

That the registrar does exercise these powers in practice is illustrated by the fact that a recent application for registration of a new union for construction workers was refused on the grounds that a construction workers' union (TAMICO) already exists. Workers can appeal through the courts but this is a lengthy and costly process. They have in the past chosen instead to apply to the Minister of Labour (in two previous cases the Minister did instruct the registrar to register the unions).

Summary of data on trade union membership

- Trade union membership is estimated to have declined from 6.4% of the total employed population in 1990 to 1.8% in 2001.
- Union membership as a proportion of paid employees was around 27% in 2000/01.
- Union membership amongst construction workers is estimated at 3.3% of the total construction workforce in 2001, falling to 1.6% in 2003.
- Union membership in Dar es Salaam in 2003 is estimated to have been 6.5% of the total employed population and 16% of paid employees.
- Union membership amongst construction workers in Dar es Salaam in 2003 is estimated at 3%.
- Union membership in the public sector in 2003 is estimated at 50% and in the private sector (excluding those employed in agriculture and the informal sector) at 14% (2003). If the informal sector is included the private sector density falls to 5%.

Once registered, trade unions enjoy considerable statutory protection, immunities and privileges. Section 47 of the Trade Union Act provides as follows:

No suit or other legal proceeding shall be maintainable in any civil court against a registered trade union or other officer or member of a trade in respect of any act done in contemplation or in furtherance of a trade dispute to which a member of the trade union is a party on the ground only that the act induces some other person to break a contract of employment, or that it is in interference with the trade, business or employment of some other person or with the right of some other person to dispose of his capital or of his labour as he wills.

Liability in tortuous acts is exempted under section 48 (10) and unions are further protected from liability in contract under the provisions of section 49.

The obligations of employers towards trade unions are also substantial. The Trade Unions Act 1998 requires employers of workers who are members of a registered trade union to deduct the union dues from the wages of the employees and pay them to the union. Further obligations for employers are outlined under section 8 of the Security of Employment Ordinance. These include the obligation to provide for the election of workers' committee members, to permit them to meet at least once a month during working hours and without loss of pay, to make a room available for meetings, to allow them to participate in inspections where the committee has such a function and to carry out other duties. The same law protects individual union leaders and committee members against discrimination by employers and specifically forbids their dismissal, unless prior approval has been obtained from the labour officer.

It is therefore evident that the existing legal framework affords workers the right to organize and join unions in workplaces and that union office bearers enjoy considerable legal protection.

Data on compliance

Trade union density is defined as the proportion of all employed persons in both public and private sectors who are members of a trade union. According to the Trade Union Congress of Tanzania (TUCTA, 2004), trade union membership before the economic reforms of the 1990s was estimated at 700,000. If this was the membership in 1990/91, it represented 6.4 % of the total employed population. Total trade union membership in 2001 was just over 300,000 (the reported figure varies from 300,747 to 313,288, averaging 307,017) and increased slightly to 317,716 in 2003. This was 1.8% of the total employed workforce of 16,914,806.

Since the vast majority of Tanzanian workers are working for themselves either in traditional agriculture or in the informal sector, a more appropriate measure of union density in the Tanzanian context would be the proportion of paid employees who are members of a trade union. There were 1,150,055 paid employees in Tanzania in 2001. Union density as a proportion of this figure would be 27%. The trade union for construction workers in the Tanzania Mines is the Construction Workers Union, or TAMICO. TAMICO membership was reported as 8595 in 2001 (TUCTA 2004). Reliable sources indicate there were slightly more construction workers than mining workers. Assuming there were 5000 construction workers in the union in 2001, that would be 3.3% of the total construction workforce. TAMICO membership fell to 6456 in 2003, of whom around 4000 were in mining, leaving only 2456 in construction. This gives a density of only 1.6%.

The director of economics and research at TUCTA confirmed that membership among construction workers has declined and attributed this to the increasing use of private contractors, who employ mainly casual workers and these are difficult to recruit into the union. Unions also have little experience of recruiting in the private sector, let alone the informal sector, having for many years relied on the public sector to provide their membership. Public sector employers even deduct union contributions from salaries at source, saving union officials the bother of having to collect union dues (TUCTA, 2004). Another major factor is a lack of manpower in TAMICO since the firing of the deputy general secretaries for construction and mining respectively. After a three year battle through the courts, the former deputy general secretary for construction was eventually reinstated, but the union remains weak.

The general situation was largely confirmed in the ILO study of large construction sites in 2003 (ILO, 2005b). A trade union presence was found on only two sites and on one of these it was said to be largely ineffective. On several sites, labourers expressed an interest in joining the union, but knowledge of TAMICO and of workers' rights in general was found to be very limited. The contractors' representatives interviewed on all sites maintained that workers are allowed to join a trade union and to hold meetings on site. However, on two sites the workers disagreed. On one of these sites the workers specifically said that they were interested in joining the union but the management would not allow it.

Compilations have been made of the numbers of trade union members in Dar es Salaam, although the data is only available for 2003. The total number of union members in the city in 2003 was 47,802 (all sectors) out of a total employed population of 738,731, or a total number of paid employees of 291,731 (2000/01 data). This is 6.5% of the total employed population and 16% of paid employees. In 2003, there were 1004 TAMICO members in the Morogoro region, which includes DSM. Even if all were construction workers (some may be mining workers) and all were in Dar es Salaam, this is still only 3% of the estimated number of construction workers in the city. It was also possible to make a rough calculation of the numbers of TU members in the public and private sectors, for the country as a whole and for Dar es Salaam. Again, the data is only available for 2003. It was found that out of the total of 316,411 trade union members, 209,996 were employed in the public sector, compared with the 106,415 in the private sector. Using 2001 employment data, this gives a public sector density of roughly 50% and a private sector density (excluding those working in agriculture and the informal sector) of 14%, but this falls to 5% if the informal sector is included. These findings are not unexpected as it is widely recognized that a high proportion of union members are working in the public sector. The decline in membership in recent years is largely a reflection of falling public sector employment. More than one-third (117,000) of all union members in 2001 (and more than half of those in the public sector) were teachers, who are employed by the Government and whose membership of the Tanzanian Teachers Union is mandatory.

Collective bargaining coverage rates

There are very few collective bargaining agreements in the construction industry, and some of them are with public sector organizations where union members are the managers and regular employees. The number of construction workers covered by such agreements is miniscule.

The TUCTA 2004 report maintains that either Government refuses to register collective bargaining agreements in the industrial court or that there are long delays in doing so. Other authors maintain that there is no collective bargaining in Tanzania.

5.4.5 Synthesis: decent work indicators in Dar es Salaam

Table 5.14 presents the decent work indicators for Tanzania and Dar es Salaam according to the four key components. The right hand column of this table shows trends towards (positive) or away from (negative) decent work for each of the indicators in all sectors and in the construction sector, at the national and local levels.

Table 5.14   Dar es Salaam decent work indicators

Employment dimension				
Unemployment rate		1990/91	2000/01	Trend towards DW
National level	All sectors	3.6%	5.1%	−ve
Local level	All sectors	n/a	26.4%	
Low wage rate		1990/91	2000/01	
National level	All sectors	8%	26%	−ve
Hours of work		1990/91	2000/01	
National	All sectors	42%	65%	−ve
Social security dimension				
Public social security coverage		1990/91	2000/01	Trend towards DW
National level	All sectors	n/a	46%	
Old age pension		1990/91	2003	
National level	All sectors	n/a	3%	
Workers' rights dimension				
Wage inequality between genders		1990/91	2000/01	Trend towards DW
National level	All sectors	120%	120%	
	Construction	125%	75%	+ve
Child labour		1990/91 (10–14)	2000/01 (10–17)	
National level	All sectors	23%	53%	−ve
	Construction	n/a	n/a	
Local level	All sectors	n/a	16%	
Social dialogue dimension				
Union density rate		1990/91	2000/01	Trend towards DW
National level	All sectors	6.4%	1.8%	−ve
	Construction	n/a	3.3%	
Local	All sectors	n/a	6.5% (2003)	
	Construction	n/a	3% (2003)	

5.5   Decent work in Dar es Salaam: best practices

5.5.1   *Initiatives of the Dar es Salaam City Council*

There are two major challenges to local government in Tanzanian cities:

- Providing basic services in planned and unplanned settlements, and ensuring that the city is a clean and healthy place to live
- Generating opportunities for productive employment and accommodating micro-enterprise activities that are not detrimental to the city environment

Some of the initiatives taken by local authorities in Dar es Salaam are outlined in this section.

Sustainable Dar es Salaam Program (SDP)

Dar es Salaam was the first city to participate in the UNCHS/Habitat Sustainable Cities Program. In the process, it shifted from a prescriptive and bureaucratic master planning tradition to more participatory and collaborative methods of working among stakeholders in an urban setting (Nnkya, 1999). This initiative originated with a request in 1990 from the Government of Tanzania through the Ministry of Lands, Housing and Urban Development to the UNDP for technical assistance to review the Dar es Salaam Master Plan. Dar es Salaam has had three master plans, in 1948, 1968 and 1979, and these were supposed to be reviewed every five years to capture social and economic changes. UNCHS at this time had just introduced its Sustainable Cities Program and was looking for cities to pilot it. Seeing Dar es Salaam as a potential pilot project, it sent an expert to discuss details with the Tanzanian Government.

The idea was not well received by the Directorate of Urban Planning. The Sustainable Cities Program centred on the relatively new idea of Environmental Planning and Management (EPM)[9]. This was an unconventional planning approach and was not expected to result in a master plan. A second objection lay in the fact that the program would be carried out by local government and not by the centre. The ministry had until that time kept control of the planning process because of inadequate capacity at the local level. However, the idea of an alternative planning approach was well received by the Dar es Salaam City Council. A project document was prepared and signed in 1991 by the Government of Tanzania, UNDP and UNCHS for what became the Sustainable Dar es Salaam Project (SDP).

The SDP established working groups to tackle key problems identified through wide consultation with communities and other stakeholders. Five problems were identified initially and this soon increased to nine. At the top of the list were inadequate solid waste management (the city was

covered with piles of rubbish) and overcrowded, unplanned and poorly
serviced settlements. In the first two years the city centre was cleared of
waste in a large-scale clean-up campaign. The various stakeholders who
participated in the campaign were the central government, the City Council, donor agencies, the private sector and individuals. Researchers have
argued that this experience led the City Council to see that improvements
in solid waste management could be realized through stakeholder participation, which set the scene for subsequent privatization (Halla & Majani,
1999; Kassim & Ali, 2006b). In 1994, Dar es Salaam City Council acknowledged the difficulty of managing solid waste on its own and decided to involve the private sector as a partner in its collection (Kassim & Ali, 2006b).
In the partnership the private sector acts as the collector (service provider),
while the City Council remains the principal, with overall responsibility
for the provision of the service and for setting the framework. This includes passing by-laws setting collection and disposal charges, enforcing
these by-laws, monitoring performance of service providers and managing the whole scheme (Halla & Majani, 1999). Multinet Africa Limited was
given the franchise to collect refuse in the ten city centre wards on behalf
of the City Council on the basis of user charges. Within six months of operation, the rate of collection had improved from about 3% city wide to
around 75% in the privatized area (Nnkya, 1999).

   While the progress made on solid waste collection and the foundation it
laid for future developments in this area (see below) was a major success
of the SDP, there were serious disappointments on the employment front.
There was disagreement between the traditional land use planners and the
environmental planning advocates as to the outputs of the project. While
some were happy with the Environmental Management Strategy for Dar
es Salaam as the basis for a flexible management of urban development,
the land use planners still required a master plan. According to one author who was closely involved in the management of the project: 'These
practitioners, comprising the law enforcers from both the central and local
governments, have regarded the EPM process ... to be wasteful in terms
of both time and other resources and therefore an interference with their
routines and bureaucratic processes of day to day city administration'
(Halla & Majani, 1999: 348).

   The effect of this can be seen in the continued activities of law enforcers.
For example, the authors cited above refer to the ongoing unilateral action of DCC to demolish kiosks used by the informal sector operators in
the name of law enforcement. They argue that this behaviour frustrates the
'opportunities and initiatives for employment creation and income generation among the unemployed in the city. The action also threatens the
commitment of other city stakeholders in resource mobilization and investment for city growth and development' (Halla & Majani, 1999: 348–
349). Writing in 1999, Nnkya concludes that 'in spite of the acceptance
by the DCC that formal and informal micro-enterprises are important

activities sustaining a significant proportion of the urban residents, appropriate strategies of how to accommodate and support these activities have yet to be developed' (Nnkya, 1999: 5).

The same author expressed concern over the neglect of economy and employment concerns in the SDP. 'Urban economy and informal sector activities were among the nine issues of concern identified by the City Consultation in 1992. For unknown reasons, right from the beginning, the SDP decided not to deal with urban economy issues, hence no working groups were established' (Nnkya, 1999: 21). The author concludes that the omission of urban economy issues from the draft Sustainable Urban Development Plan (produced to satisfy the land use planners) made it not much better than a master plan and thus, a deficient tool for guiding urban growth and development right from the beginning.

Solid waste management program (SWM)

Solid waste management is a service for which local government authorities are usually responsible. However, lack of capacity in the public sector in Dar es Salaam led to recognition in the early 1990s that the private sector needed to be involved. We have seen that the first attempts at public-private partnership to collect solid waste in the city, under the SDP, focused on the city centre and involved one private company. Although this was successful in the areas covered, still only 10% of the rubbish generated in the city was being collected. There was an urgent need to extend coverage to other parts of the city, especially to the informal settlements, where the majority of the population was living. In 1994, still under the SDP, three new working groups were established to:

- Expand privatization to new areas by involving community based organizations (CBOs) and NGOs as primary collectors
- Strengthen waste disposal site management
- Encourage waste recycling

In 1998 the ILO became involved, with funding from UNDP, in what became the Dar es Salaam Solid Waste Management Project. The project started life under the Dar es Salaam City Commission, but after 2000 responsibility was taken over by the 'solid waste management department' of the three municipalities, supported by the equivalent department in the Dar es Salaam City Council (DCC). A major objective of the SWM project was to promote employment opportunities and income-generating activities within the SWM sector, while also improving the cleanliness of the city and reducing the amount of waste by encouraging recycling (Salewi, 2006). The strategy adopted was to involve local communities through the ward and *mtaa* system (the lowest unit of local government) and to extend the award of franchises to small enterprises and community-based organizations. Each franchisee was authorized to collect waste in a specified area of the city and to collect fees directly from the local residents. During the course of the project (1998–2003) more than 50 franchises were established, covering 44 of the 74 wards in the city. The Dar es Salaam City Council supported the small enterprises involved by providing carts. DCC also managed the dump site, formulated the contracts which were let by the municipalities and coordinated activities from the grass-roots level to the final dumping of the waste. This type of collaboration between the local authorities and an incipient private sector represents a new kind of public-private partnership with enormous potential (Salewi, 2006). The result was an increase in waste collection from 10% in 1994 to 48% in 2005 (Kassim & Ali, 2006b). At the same time, more than 2000 jobs were created, 60% of them for women (Salewi, 2006).

However, a number of problems remain. Many residents in low-income areas are reluctant to pay the fees demanded for the service. The belief is widespread among the public that the service should be provided free of charge, as it was in the past. Many do not realize that the service is provided by the private sector, and those that do, believe that the service providers have already been paid by the city and are reluctant to pay again (Kassim & Ali, 2006b). This has led to a situation where many franchisees are unable to collect the fees due to them and are struggling to survive. It was assumed in the planning phase that it might be possible to cross subsidize between high- and low-income areas by making each contractor responsible for one high-income and one low-income area. This proved impossible, however, as virtually none of the contractors had the capacity to serve more than one area (Bakker *et al.*, 2000). Hence, franchisees jostle for the high-income areas and central business district where fees are greater and access easier than in the informal settlements, meaning that many low-income settlements lack franchisees to serve them (Salewi, 2006). In this context, some individuals and groups have begun to collect waste informally in the unserviced areas without a contract from the municipality.

More serious problems for the ILO are the low wages and very poor working conditions in the SWM industry. A survey undertaken in 2005 found that workers in SWM are employed on a casual daily basis and paid only Tsh1000 per day, which is well below the minimum wage (Kaseva & Mbuligwe, 2005). This was confirmed in a study commissioned by the ILO in 2003 which examined working conditions in 35 of the 44 wards that have a contract with the municipal authorities and 6 of the 29 wards where waste collectors operate informally (Kiwasila, 2003). The study found not only low wages, with 62% earning less than the minimum wage, but also wage disparity, with men earning 27% more than the women, largely resulting from their undertaking different tasks. In the study 64% of workers complained of long working hours. About 90% of workers did not have a contract but were employed on a daily basis with no job security. Although most reported a good understanding between workers and employers, absence from work due to illness, or any kind of activity to defend their right to a decent wage would result in instant dismissal. While child labour has been eliminated among the licensed franchisees, it is still present among those collecting rubbish informally as well as at the waste collection points and dumpsites (Kiwasila, 2003).

Those who work in waste collection do so under precarious conditions and workers are exposed to serious health hazards. Despite the fact that the use of protective clothing is required by the by-laws, municipal authorities do not have the manpower or the funds to enforce this. The study found the use of protective clothing to be minimal, with two- thirds of those interviewed having no protective gear. Furthermore, 84% of those interviewed reported lack of welfare assistance, even after injury at work (Kiwasila, 2003). The study concluded that much work remains to be done to promote decent work in this sector. In the words of the ILO expert responsible for a follow-up project; 'Decent work is still a goal' (Salewi, 2006: 9).

Hanna Nassif urban infrastructure upgrading project

Hanna Nassif is an unplanned settlement in the Kinondoni district of Dar es Salaam, with a population of around 20,000 located just 4 km from the city centre. The settlement lacked basic infrastructure and was also liable to flooding in the rainy season. In 1990 there were heavy rains and the community organized a brainstorming meeting to decide what to do. The priority was to reduce flooding in the area by constructing storm water drainage. The resulting ideas were taken to the ward level, then to the municipal level (Kinondoni District) and finally to the City Council. Several plans were developed but failed to be implemented due to lack of funds.

In the early 1990s, a pilot project was formulated by the ILO to tackle the problem of flooding, through the use of labour-based methods of construction involving the community in all stages of the construction process (Tournée & van Esch, 2001). In 1993 a community development committee (CDC) was formed and shared responsibility, as the contracting authority or 'client' for the project, with the DCC. The DCC also seconded staff to the technical assistance team. Work was actually implemented by the construction committee which was a sub-committee of the CDC. The construction committee became the contractor, taking full responsibility for the project's implementation, including the organization of paid and unpaid labour. Construction work began in 1994 with a grant from the Ford Foundation and finished in 1996, having completed 600 metres of main storm water drainage, 1500 meters of side drains, 1000 metres of dirt road with two protected drain outlets, improved footpaths and ten vehicular culverts (Tournée & van Esch, 2001). Construction work took longer than originally foreseen (two and a half years instead of eighteen months) but was of adequate quality, although the finishings were probably of lower quality than would have been delivered by contractors (Tournée & van Esch, 2001). The daily wage paid to the labourers at that time was Tsh700.

The second phase of the Hanna Nassif project began in 1997 and continued until 2001. The National Income Generating Program (NIGP) was the executing agent and project manager during this phase. The major objective was to expand employment opportunities for the poor living in unplanned settlements, while also improving the environment and providing access to basic services. A second objective was to build experience during project implementation in order to replicate the initiative in other unplanned areas (Salewi, 2006). Unlike the first phase, this part of the project involved the provision of a number of services, including a water supply system and a system for the collection of solid waste. It also provided support for micro-enterprise development through a community scheme for the provision of credit. However, the chief difference from phase one was that private small-scale labour-based contractors were brought in to manage the major works and only the minor works were undertaken by the community.

The rationale behind introducing contractors was to build the local community's capacity by exposing them to better project management and organization skills through employment in a contracting company. To this end, it was specified in the contracts that contractors were to use local labour from Hannah Nassif and to rotate them on a weekly or fortnightly basis to ensure that the maximum number of workers were exposed. Complaints that the contractors did not do so illustrates ongoing tension in contract documentation which 'encourages' the employment of local labour while retaining the contractors' right to have full control over recruitment and labour policies.

A further ambiguity in the objective to employ 'local labour' is illustrated by the fact that the Hanna Nassif Community Development Association (CDA), having gained considerable skill, has subsequently offered their services as a contractor in other community upgrading schemes, thereby precluding the possibility of generating employment for local labour from that community. It is suggested that there is a trade-off here between the objective of employing local labour and that of developing a specialized and experienced construction workforce. While basic construction may appear to require no skill in the formal sense, experience should not be undervalued. The attempt to retain benefits within a specific community may in the long run be counterproductive to the development of national and local construction sectors.

Phase II of the Hanna Nassif upgrading project is reported to have generated 30,483 workdays of employment, with 50% of the beneficiaries being women (Salewi, 2006). Normally, fifty unskilled and five skilled workers were employed each week. As there were many more applicants for work than jobs available, workers were rotated and applicants could only expect to work for one week out of two, three, or four, depending on the number of applicants (Clifton & van Esch, 2000). This did not pose a problem as workers were aware that the work would be only temporary, but it did lead to the perception that national legislation in this respect was irrelevant. Wages paid by community contractors were set at Tsh1600 per day for skilled workers and Tsh1200 for unskilled, which was in line with the minimum wage at the time (Clifton & van Esch, 2000) and confirmed by the chairman of the CDA. These wages were paid against set tasks and workers were free to leave once the task had been completed. It was agreed that private contractors would use the same task rates but workers were expected to continue working for the whole day with pro rata payment for the number of tasks completed (i.e. piecework). There were no records of problems with late or inadequate payment of wages. The community did complain, however, that the contractors did not teach them anything (Clifton & van Esch, 2000). Furthermore, the works executed by private contractors proved to be 30–60% more expensive than those undertaken by the community.

As this was an ILO project, the project management team ensured that the CDA and site managers were aware of national labour laws and international good practice. No under-age workers were employed and only guards worked seven days a week. Drinking water was provided at all sites and first aid facilities, workmen's compensation insurance and protective clothing were provided by the project. However, the private contractors had limited H&S equipment and relied on the CDA to supply it, and some of the work (e.g. deep excavation of unstable material on steep slopes) was inherently hazardous. The project evaluators concluded that more protective clothing should have been available and its use made mandatory.

In addition to generating 'fairly decent' employment, the project succeeded in expanding the network of gravel roads, installing a water supply system and reducing the area liable to flooding from 55% to 30%. A system of solid waste collection was also established, as well as a community fund for the provision of small loans to micro-enterprises. However, the waste collection proved to be unsustainable, running for only fifteen months, due to the fact that residents were not paying the monthly collection fee of Tsh500 per household. The micro-credit project lasted a little longer but it also eventually collapsed, as residents ceased repaying the sums they had borrowed. At the time of visiting the CDA in March 2006, the only source of revenue was from the water kiosks. The residents are charged Tsh20 per twenty-litre bucket, out of which the water kiosk attendant is paid Tsh500 per day. This sole source of revenue has to pay the water bills and running costs of the CDA office, as well as maintenance of the water supply, gravel roads and drainage system.

Maintenance is supposed to be the responsibility of the Kinondoni Municipal Council. In 2004, the CDA sent a budget for Tsh70 million (US$70,000) for maintenance of the roads to the KMC but it was not approved. The role of local government authorities in the project appears to have actually been quite minimal. The city commission was originally involved in setting standards for the project but, when the commission was dissolved in 2000, the responsibility fell upon the Kinondoni municipality. At the time of the project evaluation, the KMC declared their support, but details were not forthcoming. The evaluation report for Phase II proposed that possible support for maintenance should be discussed with the KMC. It also proposed that the KMC management committee be invited to visit the settlement 'as only a few members of the committee were familiar with the project or with Hanna Nassif'.

An interesting conclusion from the evaluation was that the 'construction of community-benefiting assets by the community (CDA) using community labour does not automatically bestow any sense of ownership or obligation to maintain those assets' (Clifton & van Esch, 2000: section 11.4). However, this is perhaps a rather harsh condemnation of the community, which needs not only the will but also the means to fund necessary maintenance work. Funds can be generated from selling the service or taxation. Residents of Hanna Nassif are unwilling or unable to pay for solid waste collection and the Kinondoni Municipal Council is unwilling or unable to provide the requested funds to maintain the roads. While residents are used to paying for water at the point of delivery, they are not used to paying for these other services. Imposing tolls for the use of the roads also proved to be impractical and the attempt was soon abandoned.

Community infrastructure program(s)

A second project to involve local communities in the provision of urban infrastructure, the community infrastructure program (CIP) of the World Bank, took off in Dar es Salaam at about the same time as the Hanna Nassif project. Both have been well documented and acclaimed as 'best practices'. However, there were some key differences. While the Hanna Nassif project relied heavily on community labour-based methods to construct gravel roads, the CIP went for a higher standard of road and used private sector consultants and contractors. Hanna Nassif was an unplanned settlement, whereas the CIP was implemented in Tabata and Kijitonyama which are planned settlements lacking basic infrastructure.

According to one respondent in the City Council, the different approach used on the two projects could be traced back to the source of funds. Hanna Nassif was financed by a grant from the Ford Foundation, whereas the CIP was funded by a loan from the World Bank. In the latter case the client had to abide by the conditions of the loan related to quality and time. Since tarmac roads were specified by the donor, this ruled out the use of labour-based methods and the need for machinery ruled out reliance on community labour. Nevertheless, the CIP was recognized as one of ten best practices worldwide under UNCHS/Habitat's 'best practices and local leadership program'. The recognition was for community involvement in the planning, and notably the financing, of infrastructure in the 'challenging context of sub-Saharan Africa' (Seragelden *et al.*, 2000: 8–9), as opposed to employment generation, which (according to one informant) was not a priority of the project.

A community development officer at the Temeke District Council provided further information on the project in Tabata. The community contributed 5% towards the total cost of construction, which amounted to Tsh17,000 for each house. Construction started in 1997/98 and ended in 2000. The two contractors engaged were Japanese and Chinese. Contractors were required to use labour from the community and expected to pay market rates. Personal protective equipment (PPE) was provided by the Japanese company, but the Chinese, who are known to have poor standards of H&S, provided none. The Chinese company also paid very low wages, which led to materials being stolen by the workers to compensate for the low pay. As a result, the specified quality was not achieved. The municipality is expected to meet the cost of maintaining the roads built under the CIP project.

The Community Infrastructure Upgrading Project (CIUP) is now trying to replicate the CIP in other communities in Dar es Salaam. This project, which started in April 2005, is also funded by the World Bank as part of the Local Government Support Program. The program is demand driven and community led. Communities at ward or *mtaa* level (the lowest level of local government) decide the infrastructure they need and contribute in kind through the provision of free labour. Municipal councils also have to contribute 5% in cash and sometimes the communities help to raise this. Funds and assistance are channelled through three municipalities. Kinondoni has six projects, as does Temeke and Ilala has four. A total of 380,000 households are to be reached.

The project manager insisted that all activities funded under the project use labour-based methods and local manpower to create employment for the local population. When we asked about working conditions, however, we were told that is not the concern of the World Bank. The coordinator of the project at Dar es Salaam City Council agreed that there were no special requirements for the terms of employment or conditions of work in the contract, as that is the contractors' business. She explained that the trunk roads are tarmac and constructed by contractors. It is also part of the agreement that contractors will use local labour for minor works and ancillary tasks, and they do so. She felt that employment is important but expressed the view that this will be generated automatically during construction and, after construction is finished, through the growth of new businesses. Hence, the Council does nothing extra to create employment.

Other initiatives of the City Council

An interview with the chief engineer of Dar es Salaam City Council revealed details of a project to provide plots for house construction in new planned settlements on the outskirts of the city. The client is the Ministry of Lands and DCC is acting as consultant. The project has been running for four years and will continue into the future as the money raised from the sale of plots is reinvested in order to provide additional plots, for which there is a very large demand. The roads are constructed prior to allocating the plots, but the services are expected to come later. So far, 450 km of roads have been constructed. The roads are gravel and construction contracts are let in many small packages so as to be accessible to small local contractors. Construction techniques are mixed; partly labour-based (e.g. the trenches) but some machinery is also employed. Although employment creation is not the main objective of this project, the city engineer pointed out that it is in fact creating a lot of employment, both in the initial road construction and also later as plot owners begin to build their houses. Asked about the quality of employment, he replied that minimum wages are required in the contract and 'as all workers in Dar es Salaam know what the minimum wage is, there is no reason for them to work for less'.

Interviews with Dar es Salaam City Council officials revealed several other initiatives to create employment, although not in the construction sector and not directly. In fact, the prevailing view expressed by all officials interviewed is that employment creation is now the responsibility of the private sector. This is in line with the Government's employment policy and means, in practice, that most people will have to create jobs for themselves. The DCC does what it can to support private efforts. For example, 10% of the revenue of the city council and the three municipalities is set aside to assist petty traders through a community bank. An industrial park has been created with production and selling areas to help the *machingas* (street traders). All three municipalities are also providing spaces and creating facilities for these activities. In sum, the council's core activity is now 'enabling'.

In fulfilling their obligations in this respect, council officials have to work with other NGOs, CBO and international organizations. Sometimes the initiative may come from outside but council officers also come up with their own ideas and look for external funding to implement them. One instance led to Dar es Salaam's participation in UN Habitat's 'safer cities' initiative. The aim of the project was capacity building for crime prevention but activities also included action to address the underlying causes of crime; the most important of which was joblessness, particularly among the youth of the city. When the project started in 1998, youth groups were already undertaking night watches of their neighbourhoods with contributions (torches, clubs, etc.) from the community. These were very effective in diminishing crime, but as it reduced the groups tended to break up, which is precisely when income-generating projects are needed. The project has provided seed money for activities such as car washing, gardening, dressmaking, food processing, carpentry, poultry raising and a nursery school for the children so that the women can work. However, when we asked about the 'quality' of jobs, stressing regular wages with benefits, the response was 'such jobs are hard to come by'.

Another initiative that came from a council official but was subsequently supported by outside funds (in this case from UNEP and the World Bank) is the Dar es Salaam Rapid Transit (DART) project. It aims to greatly improve public transport in the city, with the introduction of new buses and cycle routes. The initiator of the project explained the importance of getting local support if a project of this kind is to be successful and sustainable. A financial contribution from the Council and/or the Government is an important indicator of support, but on its own is not enough. Seminars, publicity, etc., are also needed to convince politicians and the public. The DCC is now committed to DART and has voted it the number one priority of the council – ahead of street lighting and the construction of an abattoir. Evidence of the Government's commitment is seen in the project's inclusion in the recent election manifesto. This project is not specifically about employment creation but the example does illustrate how local councils can play a leading role, even when heavily reliant on external funds.

5.6 Synthesis: decent work, evidence, obstacles and potentials

5.6.1 Evidence

The labour force survey reveals that over 80% of the economically active population in Tanzania is working in traditional agriculture. More than half of the rest earn a living in the informal sector. There was little change in these proportions between 1990/91 and 2000/01. So only a minority of workers (around 10% of the working population) are in paid employment in the 'formal' part of the economy.

Even for this minority, the terms under which they are employed and the conditions in which they work would seem to fall very short of the ILO definition of 'decent work'. A few do enjoy stable employment with some degree of security and ability to exercise their rights but, with the decline in public sector employment and the private sector's preference for employing casual labour or outsourcing labour requirements, the number is small and falling. It is estimated that in 2000/01 only 4% of the total working population were regular employees in formal establishments.

In the construction industry, survey data for 1990/91 and 2000/01 shows a quite dramatic shift in the structure of employment, with a significant fall in the proportion of paid employees, from 70% to 37%, and a corresponding increase in the self-employed, from 22% to 60%. By 2000/01, the industry was characterized by very high levels of self-employment and similarly high levels of employment in small enterprises with no legal status (the informal sector). The Employment and Earnings Survey of 2000/01 further reveals that, amongst those employed in formal (registered) private enterprises, the proportion of casuals was very high (47%

compared with an average of 15% for all sectors), while the average wage for construction workers employed as casuals was less than half that paid to regular workers. Field surveys reveal an even higher proportion of casuals, not less than 70% of the total workforce and over 90% on some sites. Casual workers do not enjoy the benefits of those in regular employment, such as paid leave, weekly rest days, accident insurance or pension provision and in some instances are even denied basic welfare provisions and protective clothing. No casual construction workers are members of a trade union.

Unfortunately, there is no comparable data on the extent of casual labour in earlier years. Comparable data at two time periods (1990/91 and 2000/01) is only available for a few of the decent work indicators. However, where data is available, there is clear evidence of a trend of deterioration in working conditions. Unemployment in all industries at the national level rose from 3.6% in 1990/91 to 5.1% in 2000/01, and in urban areas from 10.6% to 14.8%. The proportion of paid employees earning less than half of the median wage rose from 8% to 26%. Trade union density fell from 6.4% to 1.8%. The proportion of paid employees working more than 49 hours per week rose from 42% to 65%. In the construction industry, usual working hours rose from 51 to 62 in the ten-year period.

The construction sector also experiences a relatively high incidence of accidents and work-related ill health. The informal sector survey of 1995 reported 13 accidents per 100 workers in construction, compared with five for all sectors combined. A survey of 63 large building sites found an average of three accidents per site each year. The high accident rate is particularly serious as only those few workers in regular employment in formal enterprises may be covered by formal insurance schemes and receive incapacity benefit if they are injured or fall sick.

Dar es Salaam city has more than its share of paid employees and construction workers. With 7.5% of the population of Tanzania, the city accounts for 25% of paid employees and 30% of construction workers in the country as a whole. Nevertheless, at least half of the employed population of Dar es Salaam and two thirds of construction workers in the city are employed in the informal sector. Many of these are in fact unemployed for much of the time. With rapid growth from in-migration, the unemployment rate in the city is extremely high, estimated in 2000/01 at 26% by the relaxed standard definition, or 46.5% by the Tanzanian definition, which includes those who are only marginally attached to work. There is no comparable unemployment data for Dar es Salaam in 1990/91.

In this context, the challenge for local authorities to create employment opportunities in the construction sector with decent working conditions is clearly enormous. It is complicated further by an ongoing and critical shortage of revenue in local councils and their continued dependence on central government or outside assistance for capital investment funds. However, despite these obvious difficulties, some things have been achieved that might point the way to greater success in the future.

Perhaps because of the lack of capital and the obvious need to depend on outside help, Dar es Salaam City Council has been very adept at taking advantage of international initiatives. This was particularly true during the period when council officials had greater freedom to make decisions outside of council control. The Sustainable Dar es Salaam Project was the first substantive involvement of UN agencies with the local council. Although the project evaded the difficult issue of employment generation, it did set a precedent for the involvement of communities and other stakeholders, including the private sector. The participatory approach that evolved to solve the problem of the accumulation of solid waste in the city set the scene for the subsequent development of Tanzanian forms of public-private partnerships. Other innovative projects followed, with outside assistance, notably projects involving local communities and providing jobs for local labour while upgrading infrastructure in Hanna Nassif, Tabata and Kijitonyma.

5.6.2 Obstacles

Finances

Finances are clearly still a major problem, limiting the ability of all four councils in Dar es Salaam to invest in new infrastructure projects and to maintain the infrastructure that already exists. Transfers from central government have always been the main source of funds for local government in Tanzania, accounting for as much as 90% of local government expenditure[10]. Prior to 2004, there were numerous shortcomings in the system for making such transfers, resulting in an inequitable, non-transparent and inefficient allocation of resources. The allocation of funds from the development budget to local government authorities over the years has been small and highly irregular. Only 4% was allocated to regions and local government authorities in 2003/04[11].

However, the whole system of government transfers is currently being transformed. From 2004/05, formula-based allocations were introduced for recurrent expenditure on primary education and health care and for other priority sectors in 2005/06. The allocation formula for local roads is based on the length of the road network (75%), the land area (15%) and the number of poor residents (10%). The Government is now also introducing a formula-based system of local government capital development grants to councils based on their performance (URT, 2005b). The scheme is funded by a group of donors in collaboration with the World Bank. In the first release of funds in May 2005, the Ilala Municipal Council received Tsh374 million and the Temeke Municipal Council received Tsh454 million[12]. Kinondoni apparently received nothing, which could be due to poor performance. Fifty per cent of the grants allocated to each district or municipality will be spent on priorities determined at the village or ward level.

While recognizing that funds transferred from central government will continue to be important and are an appropriate way to finance services agreed upon at the national level but provided by local government ('devolution of provision'), there is general agreement that other services which are truly local (or 'fully devolved') should be financed from local revenues, so as to maintain the link between costs and benefits[13]. Yet, the opportunities for local government in Tanzania to raise revenue for local priorities have recently been severely reduced. The Local Government Finance Act of 1982 took a permissive approach to local taxation, empowering local authorities to raise revenues from taxes, licences, fees, charges, etc., more or less as they wished. As a result, there was a proliferation of local taxes, not all of which were justified in the eyes of the public. The result was widespread resistance to paying, resulting in low returns. The main source of local revenue for many years was the development levy, a kind of head tax paid by all adults. It was extremely unpopular and was abolished in 2003. Many other 'nuisance taxes' have also been removed. The main source of local revenue is currently property tax.

The problem of public resistance to paying local taxes persists. Tax evasion is widespread, as is non-payment of fees and charges. A survey of 1260 citizens in six district and municipal councils in 2003 (including Ilala in Dar es Salaam) found three dimensions of trust affecting citizens' compliance:

- Trust in local government to use the revenues to provide the expected services
- Trust in the authorities to establish fair procedures for revenue collection
- Trust in fellow citizens to pay their share (Fjeldstad, 2004)

The first of these three dimensions is perhaps the most significant. The perception that revenues are not spent on public services is pervasive, reflecting a deep mistrust of local authorities' ability and motivation to provide services. People are reluctant to report misuse as they consider that all civil servants are corrupt and protect each other. They also fear repercussions. The problem may have been exacerbated by the fact that local revenues have had to be spent on local administration due to inadequate transfers from central government, which increases the public's scepticism that they will benefit from paying local taxes. It has been argued that an unconditional grant to fund core administrative functions at the local level would allow locally collected resources to be used for the delivery of services which visibly and directly provide benefits to the local community[14].

An alternative which may be preferable to providing services out of general taxation is to require direct payment for specific services, thereby strengthening the link between demand and supply. Refuse collection provides a good example of this principle. However, some services (e.g. roads and drainage) are not easily financed through cost-recovery charging systems. The solution put forward for financing these kinds of services is that the community make contributions as a whole. Contributions may be in cash or in kind in the form of free labour, but both are problematic.

Corruption

Corruption deserves special mention in this context. We have seen that fear of corruption amongst local officials is a major factor behind the public's reluctance to pay taxes and hence, a limitation on the councils' ability to collect revenue to fund projects. These fears are not unfounded. An official investigation into corruption found that local council leaders receive bribes to facilitate the award of tenders, allocation of plots and marketing stalls, and in the procurement of goods and services, particularly roadworks[15]. Problems identified in local government procurement include non-adherence to procedures, lack of transparency in prequalification and contract award, political interference, conflict of interest, incompetence and collusion with vendors. Anecdotal evidence uncovered during the ILO research with informal construction workers suggests that straightforward bribery to obtain contracts or work is commonplace at the community level.

The Warioba Report made a number of recommendations regarding procurement, which have been largely addressed by subsequent legislation and changes in the institutional framework. However, many feel that the situation on the ground has not improved significantly. In opinion polls conducted for the annual report on the state of corruption in Tanzania, one-third of citizens reported concrete experience of corruption in local authorities, particularly in the area of procurement (Anon, 2002).

> Councillors often flout tender regulations. Sometimes district engineers collude with contractors and some councillors to flout tender regulations for personal benefit. At other times, however, they are merely victims of selecting contractors for road works in the Finance and Planning Committee which can deliberately ignore technical advice on tender evaluation and award contracts to undeserving bidders. These practices often result in inflated costs, poor quality goods and services and delayed work completion.   (Anon, 2002: 84)

The implications of this kind of corruption for employment generation are extremely serious. The immediate effect is that inflated costs on any project serve to reduce the quantity of services, and hence of employment, that can be delivered for a given sum. It also seems doubtful that contractors and officials who can inflate costs for their own purposes will make any serious effort to promote decent wages and conditions for construction workers. However, the more pernicious and long-lasting effect is the erosion of urban residents' confidence in their governance institutions (Ngware, 1999).

Contracting out

A further obstacle to the creation of 'decent work' lies in the current practice of contracting out services previously provided by the Council to the private sector. A number of the council officials interviewed referred to the dramatic change that has occurred in the way in which services are procured. In the recent past, Dar es Salaam City Council had a significant manual labour force and constructed a number of projects using direct labour or 'force account'. Significant examples are the Benjamin Mkapa school and the DCC's own office in the centre of the city, which was completed in the late 1990s. Maintenance work was also undertaken by the council's own workers. Today the council contracts out almost all services and the change is recent. In 2003 there were 12 painters employed directly in the building section of the DCC. Today there are only three. In March 2006, there were 16 manual workers employed at the Karimjee Hall (former Parliament building), tending the grounds and looking after the building, but this service was later contracted out at the start of the following financial year.

We have seen that the public sector, including local authorities, has good terms and conditions for salaried employees and that casuals employed by the Council are always paid the minimum wage or above. When services are contracted out to the private sector, however, there is in practice little control over the terms and conditions of employment, including the wages paid. Contracting out to the private sector implies that the terms and conditions of employment will be set by the market. While there may be some obligations laid out in the contracts, for example to respect the minimum wage, this cannot be confirmed. In any case, there is no evident means of either monitoring to see if these obligations are respected or enforcing the terms and conditions of the contract. Many of those interviewed take the view that employment conditions are the business of the employer and ensuring that central government labour regulations are observed is the responsibility of the labour department. Others said that they preferred force account, but this is not in line with current thinking and policy in Tanzania is dictated from outside.

5.6.3   *Potentials*

In sum, the case study of Dar es Salaam has shown that, even in the absence of funds for capital expenditure, employment can be generated for the provision of services through partnership with the private and/or community sector. Nonetheless, the service provided must be something for which the public is prepared to pay directly, which is more likely to be achieved in high than in low-income areas. Solid waste collection falls into this category. It is possible that other such services could be identified which may be provided with minimal capital input or by engaging in partnership with private investors. Providing these services to low-income neighbourhoods where the need is greatest, however, will probably require cross-subsidization.

To provide employment in construction, as opposed to related services, capital for investment is required. For ongoing employment, a source of revenue for maintenance must also be identified. Recommendations may be summarized as follows:

- The potential for employment through the provision of services for which the public is prepared to pay directly should be fully exploited.
- The objective of creating employment for local labour should be balanced with that of developing a specialized and experienced construction labour force.
- Nothing should be constructed that cannot be maintained and for which a revenue stream for maintenance cannot be identified.
- If the desire to improve employment conditions is serious, then more labour should be employed directly by urban authorities or local communities. The only way in which conditions of employment will improve is by central government strengthening its enforcement of labour legislation.

Notes

Chapter opening image: Construction workers pointing a brick wall in Dar es Salaam, Tanzania. Photograph courtesy of Jill Wells.

(1) World Bank Country Brief, May 2006.
(2) This age was chosen as the cut-off point so as to make the survey comparable with that of 1990/91.
(3) The cut-off point for inclusion in the informal sector n 1990/91 was employment with fewer than five employees and in 2000/01 this was raised to ten. Also in 2000/01, professionals who fulfilled other criteria were included in informal sector activities.
(4) Seven civil engineering projects were included in the research: Makuyuni – Ngorongoro Road (Japanese contractor, Tanzanian subcontractors), Chalinze – Melela Road (Danish contractor, Tanzanian subcontractors), Songwe – Tunduma Road (Chinese contractor, Tanzanian subcontractor), Nangurukuru – Mbwemkuru Road (Chinese contractor), Mbwemkuru – Mingoyo Road (Kuwaiti

contractor, South African subcontractors), Somanga – Matandu Road (Chinese contractor) and Songo Songo Gas Development (Indian contractor, Tanzanian and foreign subcontractors). Four building engineering projects, all of them in Dar es Salaam, were also included in the study: National Housing Office Complex (Chinese contractor), Boko Housing Complex (National Housing Corporation and local labour), Bank of Tanzania Extension (South African contractor and subcontractors) and Primary School Facilities (Japanese contractor and Tanzanian subcontractor). All but two of the projects are large by Tanzanian standards, ranging from US$10 million to US$97 million.

(5)   'Taking stock of the state of local government finances and the policy debate on local government finances in Tanzania', anonymous report probably authored by the World Bank (May 2006). www.tanzaniagateway.org

(6)   It should be noted that there was also an informal sector survey in 1991, which covered the country as a whole, but the data is not strictly comparable with that of 1995, as the cut-off point for inclusion was five employees in 1991 and ten in 1995.

(7)   In 2002, Tsh1000 were worth approximately US$1. Sheuya (1996) gives the rate in 1996 as Tsh600 to US$1.

(8)   In March 2006 there were reports in the newspapers of NSSF investing Tsh45 billion to buy an estate of warehouses in Dar es Salaam that it had previously loaned the builder Tsh9 billion to construct.

(9)   The underlying concept of the EPM process was developed and adopted through the Urban Management Program, an inter-agency facility between UNCHS (now UN-Habitat), UNDP and the World Bank.

(10) 'Taking stock of the state of local government finances and the policy debate on local government finances in Tanzania', anonymous report (probably authored by the World Bank) posted on the Government website www.tanzaniagateway.org

(11) *The Guardian*, Friday 24 March 2006.

(12) *The Guardian*, Friday 24 March 2006.

(13) *The Guardian*, Friday 24 March 2006.

(14) *The Guardian*, Friday 24 March 2006.

(15) *State of Corruption in the Country*, Report of the Commission chaired by Hon. J. Warioba, Dar es Salaam, December 1996.

6 Santo André

Mariana Paredes Gil

6.1   Introduction

This chapter presents the case study of Santo André. The chapter begins with an overview of the political context and examines economic development at both the national and local levels. It then determines the local employment situation using some specific indicators of the four components of decent work. Available information, including official statistics, has been analyzed for the years 1992 and 2001. Qualitative information, mainly derived from interviews with key actors of different institutions in Santo André that are involved with the construction sector, has been considered. The fieldwork also examined efforts made by the local authority to promote employment in the construction sector and related services, and to assess how the municipality has been able to apply the principles of decent work through examples of good practice, such as the Santo André Mais Igual (SAMI) Program.

6.2   National, regional and local context

6.2.1   National context

Brazil is a federative republic consisting of the Federal District (Brasília) and 26 states. Each state possesses its own legislative assembly and state governor who are elected by popular vote. Each state enjoys significant autonomy as far as government, legislation, public security and taxation are concerned. Furthermore, the municipalities enjoy a degree of autonomy from both federal and state governments. Figure 6.1 shows the main cities in Brazil.

Political situation

The Federal Republic of Brazil was declared in 1891, with the promulgation of a constitution which lasted until 1930, when economic depression and regional disagreements provoked a military coup that brought Getúlio Vargas to power. The Vargas Government, and the military leaders that supported him, focused economic policy on industrial development. However, growing economic inequities and the intensification of agitations led to another military coup in 1964. The next government (1964–1967) was characterized by its policy favouring economic austerity, and gave way to the heterodox policies that gave rise to the Brazilian economic boom between 1968 and 1973: high rates of growth, together with industrial diversification and trade liberalization. However, this liberalization was not followed by reforms in the political area. Despite the economic successes of the 'Brazilian miracle', the country accumulated large external deficits, and was dependent on foreign capital inflows. These weaknesses became more evident after the oil crisis of 1973.

Figure 6.1   Brazil's main cities
Source: http://geography.about.com/library/cia/blcbrazil.htm

With a deteriorating economic situation and increasing popular unrest, the military Government adopted a program of progressive political liberalization. By 1985, this process had increased to the point where the military Government permitted an indirect civilian election for presidency. Tancreto Neves (the candidate of the Partido Movimento Democrático Brasileiro, PMDB) won, but died on the eve of his inauguration. His Vice-President, José Sarney (a member of the conservative Partido da Frente Liberal, PFL), became President. He was the first civilian to hold that office in more than 20 years. At the beginning, Sarney's government was relatively successful, but his reputation was damaged when his controversial economic stabilization package, the Cruzado Plan, collapsed in 1987. This paved the way for the success of a political outsider, Fernando Collor de Mello, in the 1989 presidential election. Collor succeeded in the implementation of a policy of trade and market liberalization, but failed in his attempts to restore price stability. In 1992, accusations of corruption against the President and his political associates resulted in Collor's impeachment by the congress. He was replaced by the Vice-President, Itamar Franco, who served for the rest of the Government's term, but did not do much to address the hyperinflation.

The economic situation began to improve after Fernando Henrique Cardoso, a former left-wing academic turned senator, was designated

Finance Minister in May 1993. His economic stabilization plan – the Real Plan – was introduced in December 1993 and concluded in July 1994 with the launch of a new currency – the Real – which had its exchange rate linked to the US dollar. The plan's success, which rapidly ended hyperinflation and boosted real incomes, swept Cardoso to the presidency with 54% of the vote in the first round of the national election in 1994. Cardoso won congressional approval for legislation to deregulate the economy and open it up to foreign capital. He also began to implement a successful program of health care and primary education reforms. The combination of the opening of the economy, a progressively overvalued exchange rate and the lack of progress on fiscal reform ended in external and fiscal financing requirements that left the country exposed to contagion from the Asian and Russian crises of 1997–1998. In this context, Cardoso was elected to a second term as President in October 1998.

In 2002, in a situation of fiscal strictness and low economic growth, Luis Ignacio Lula da Silva, the candidate for the left-wing PT (Partido dos Trabalhadores), almost won the first round of the presidential election. In the second round in October 2002, da Silva won the election, gaining 61.3% of the vote and defeating the centrist candidate José Serra of the Brazilian Social Democracy Party (Partido da Social Democracia Brasileira, PSDB). It was a historic victory for the PT, which became Brazil's first left-wing government. Da Silva's first term of office was strongly criticized, although all expectations were exceeded in terms of macro-economic management. Even though the da Silva administration had made inroads into reducing poverty and unemployment had declined, the levels of both remained relatively high. Until mid-2005, da Silva's popularity remained high, despite these social disappointments. Nevertheless, a series of corruption scandals within the PT were threatening. All these allegations could not impede the re-election of Lula da Silva for a second term, however, and in October 2006, he won the second round by a substantial margin.

In Brazil, the promotion of decent work builds upon a political commitment between the Brazilian Government and the ILO. In June 2006, the President of the Brazilian Republic, Luis Ignacio Lula da Silva, and the ILO Director-General, Juan Somavia, signed a Memorandum of Understanding that provides the foundation for a Special Program of Technical Cooperation for the Promotion of a National Agenda on Decent Work. The Memorandum of Understanding established four priority areas of co-operation:

(1) Employment generation, micro-financing and human resources training;
(2) Viability and extension of the social security system;
(3) Strengthening of tripartism and social dialogue;
(4) Fight against child labour and the sexual exploitation of children and adolescents, forced labour and discrimination in employment.

Economy

Brazil has comparative advantages in the sectors of agriculture and pro-cessing of primary goods owing to its vast natural resources. In 2000, agri-culture represented around 10% of the GDP, with important agricultural commodities such as coffee, soybeans, sugar, tobacco, cocoa, oranges and meat. Other important industries for processing primary goods include leather footwear, wood products and mineral and metal products (iron, steel and aluminium). In 2000, industry accounted for 40% of the GDP. Brazil's extensive and diversified industrial base, ranging from heavy en-gineering to consumer goods, was largely developed in response to a Government policy: import-substituting industrialization (ISI). This pol-icy lasted 35 years and ended in the 1980s. The Economist Intelligence Unit (2005: 24–25) noted that the success of this policy, which involved trade protection plus direct support for industrial investment, was facili-tated by the huge domestic market. The services sector, which ranges from unskilled and low value-added personal services to high-earning profes-sional and financial services, represented 50% of the Brazilian GDP in 2000. Brazil has traditionally been a relatively closed economy, an outline of the import-substituting industrialization development model. However, the trade liberalization that began in 1990 has brought major changes. The elimination of the vast majority of non-tariff barriers and the slash-ing of imports tariffs strengthened Brazil's comparative advantage in agri-culture, while highlighting the competitive disadvantages of other sectors (EIU, 2005: 24–25).

A substantial wave of industrialization took place in Brazil between 1930 and 1980 – 'import substitution industrialization' – which largely succeeded in providing a significant boost to the Brazilian economy. As a result, many major Brazilian and foreign companies established indus-trial businesses in specific areas, such as the ABC region, with a view to exploiting new opportunities in world markets while contributing to the acceleration of Brazil's economic growth.

Although by the end of the 1970s Brazil had one of the developing world's largest industrial complexes, the tragic consequences of acceler-ated urbanization and an unbalanced development model became evident in a range of negative social indicators, such as high concentrations of in-come and wealth, low educational levels, precarious housing conditions and limited access to public services for the majority of the population. Difficulties in meeting external liabilities during a period of restructur-ing of the world economy led to high inflation, economic stagnation and higher unemployment in all sectors though the mid-1990s.

From 1990, the national economy endured a series of major changes, with a rapid opening up of the economy to international trade and com-mercial integration within the MERCOSUR (the Southern Cone Common Market with Argentina, Paraguay and Uruguay). Brazil returned to in-ternational financial markets. It increased the innovation of products and processes (technological, organizational and managerial) and privatized state-owned companies. Even if these changes targeted national growth, certain negative effects of the renewed economic activity had a serious negative impact on labour conditions, specifically on access to jobs. Commercial 'openings' provided the opportunity for modernizing the manufacturing sector, with lower costs and higher quality. The larger companies faced a degree of interruption in their respective manufacturing chains together with labour costs and logistical changes, leading to the rapid dispersal of some of the industrial complex, with many larger industries moving to areas where lower costs were more than compensated by the distances involved. This was primarily the case for the automobile parts industry in the ABC Region[1]. From the early 1990s onwards, despite the continuing existence of long-established industrial concerns and the strong presence of major companies, the ABC economy suffered as a consequence of the overall restructuring process, which accelerated industrial decline and led to employment cuts.

Population

In 2000, according to the Brazilian Institute of Geography and Statistics (IBGE), 81.2% of the Brazilian population was urban, up from 78.4% in 1996 and 75.6 % in 1991. The trend towards urbanization has been motivated by three factors: higher rates of population growth in towns and cities, strong migration flows from rural to urban areas and the urbanization of areas previously categorized as rural. During the 1990s, the growth of the urban population was been strongest in the northern, north-eastern and central-western regions.

By international standards, Brazil's population remains very young. Nevertheless, since the beginning of the 1990s, the percentage of persons aged less than 14 years has decreased, while that of those aged 65 and over has increased. The participation of people over 60 years of age in the workforce is also rising.

Despite Brazil being a large country with abundant resources and a significant economy, over 22 million people live in extreme poverty. If people in relative poverty are included, the total number – 53 million, or approximately 30% of the total population – live with incomes that fail to meet their basic needs. This is shown in Table 6.1, which depicts the income distribution pattern in the country.

The construction sector

At the national level, the construction sector is perceived as an important component of the economy. According to the Ministry of Development, Industry and International Trade (MDIC), in the year 2000 the sector accounted for 18% of GDP. This figure incorporates building materials production and all services related to the post-construction stage,

Table 6.1   Income distribution in Brazil (1981–2003)

Percentage of population	1981	1990	1995	1998	2001	2003
1% wealthiest	13.0	14.6	13.9	13.7	13.6	13.2
5% wealthiest	33.4	35.8	34.6	34.0	33.7	33.1
10% wealthiest	46.6	49.7	48.2	47.5	46.9	46.1
90% poorest	53.4	50.3	51.8	52.5	53.1	53.9
50% poorest	13.4	11.2	13.0	13.5	14.4	14.8
10% poorest	0.9	0.8	1.1	1.2	1.0	0.7

Source: IBGE (2003) *Pesquisa Nacional por Amostra de Domicílios.*

such as real estate activities. In 2000, according to the National Accounts System[2] (IBGE, 2005), building and heavy construction accounted for 8.7% of GDP, as shown in Table 6.2. From 1990 to 2003, the construction sector was responsible for 60% of all gross capital formation, with machinery and equipment representing around 30%. This has declined since 1998, together with a declining share of the construction sector in GDP.

According to DIESSE (2004: 5), and as shown in Table 6.3, the construction sector in Brazil employs around 6.5% of the occupied population, and two main segments absorb the majority of workers: building works (involving 70% of workers) and infrastructure sites (12% of workers).

During the 1990s, the composition of workers in the construction sector changed significantly. As shown in Table 6.4, an important general trend was the increasing number of women working in the construction sector, a trend which has occurred in almost all the different functions. Another significant trend has been the increase of the self-employed and employers in the construction sector, while employees and self-build construction workers have decreased, as have non-remunerated workers.

The majority of workers in the construction sectors do not have any education or formal training. They are trained during their work, without any

Table 6.2   Participation of economic sectors in GDP, 1991–2004

Sector	1991	1995	1998	2000	2004
Processing	24.9%	23.9%	20.7%	22.4%	24%
Other services	12.1%	12.1%	12 %	11.3%	10.5%
Agribusiness	7.8%	9%	8.2%	8%	10.1%
Real estate	12.8%	10.4%	15%	12.7%	9.4%
Commerce	9.8%	8.9%	7.1%	7.4%	7.8%
Construction	7.1%	9.2%	9.6%	8.7 %	6.8%
Financial	13.9%	8%	6.5%	5.4%	6.6%
Minerals	1.6%	0.9%	0.6%	2.6%	4.2%
Public services	2.6%	2.7%	3.2%	3.5%	3.4%
Communication	1.2%	1.5%	2.6%	2.7%	3.1%
Transport	3.8%	3.4%	3%	2.7%	2.2%

Sources: DIESSE (2004) *Caracterização do Setor da Construção Civil*; IBGE (2005) *Pesquisa Nacional por Amostra de Domicílios.*

Table 6.3 Distribution of occupied people by economic sector, 2003

Sector	Number	Percentage
Agriculture	16,409,383	20.7
Industry	11,387,016	14.4
Construction	5,157,554	6.5
Domestic trade	14,047,477	17.7
Accommodation and alimentation	2,858,332	3.6
Transport and communication	3,680,609	4.6
Public administration	3,942,196	5
Education, health and social services	7,087,297	8.9
Domestic services	6,081,879	7.7
Other services	2,947,023	3.7
Other activities	5,455,622	6.9
No declared activities	196,239	0.2
Total	79,250,627	100

Source: DIESSE (2004) Caracterização do Setor da Construção Civil.

type of previous qualification. Werna (1997: 17) notes that 'the common path for a worker is to enter the industry as an apprentice (servente), then build his way up through on-the-job training to the position of a craftsman (official) (such as bricklayers, electricians, etc.). A small minority reach a further stage of foreman (mestre-de-obra).'Werna also notes that, according to the Brazilian classification, 'the apprentice carries out simple manual tasks, such as digging holes, transporting and mixing building materials, setting or dismantling wooden frames for concrete, among others. The craftsman carries out the actual building work, helped by the apprentice. The foreman organizes and supervises the activities of the two types of labourers noted above.'

Jófilo Moreira Lima Jr. et al. (2004) carried out some interesting research on the profile of the labour force in the construction industry. Some of the key findings are:

- Low qualifications: 72% of workers have never attended any training programs at all, 80% have not finished elementary school and 20% are illiterate;

Table 6.4 Personnel working in civil construction by job status and sex (percentage) 1992–2003

	Employee		Self-employed		Employers		Non-remunerated		Self-build construction workers	
Year	Male	Female	Male	Female	Male	Female	Male	Female	Male	Female
1992	55.2	65.6	37	3	3	0.6	1.6	7.1	3.3	23.6
1997	50.4	63.1	40.8	3.1	3.9	6.7	1.2	4.2	3.7	23
2002	48.3	67.1	44.2	7.5	4.1	8.4	0.9	6.1	2.4	10.9

Source: IBGE (2005) Pesquisa Nacional por Amostra de Domicílios.

Table 6.5   People occupied in the construction sector in Brazil, 2003

Type of workers	Number	Percentage
Total occupied construction sector	5,157,554	100
Independent workers	2,345,730	45.5
Registered employees	1,044,859	20.2
Non-registered employees	1,372,594	26.6
Work more than 44 hours/day	2,493,905	48.4
No contribution to social security	3,715,181	72
Contribution to social security	1,442,373	28

Source: DIESSE (2004) *Caracterização do Setor da Construção Civil.*

- High turnover: 56.5% stay less than one year in a job and 47% work in the sector for under five years;
- Low wages: 50% of workers earn less than two minimum wages;
- High level of absenteeism (this is due mainly to health problems): 52% had at least one absence in the month prior to the study; 14.6% of workers suffered some sort of accident at work during the year prior to our research, corresponding to 148,000 people;
- Alcoholism: 54.3% are users of alcoholic beverages, 15% are excessive users, and 4.4% are alcohol-dependent.

Table 6.5 shows the distribution of people occupied in the construction sector in 2003. A large number of workers (45.5%) are qualified as independent. No less than 26.6% of workers in the construction sector are not registered. Consequently, 72% of them do not contribute to any social security system.

DIESSE (2001: 10–20) noted that since the 1990s, the construction sector has been modified following a restructuring process and the implementation of innovations that have effects on daily work. Most of the time, these innovations concern two main elements in the productive process: the utilization of new technologies and new forms of labour force management. In parallel, a new conception of the construction sector has arisen: the use of prefabricated systems. Today, entire stages of the construction process no longer occur on construction sites. These changes have important consequences for workers. A whole new series of demands, including new and better qualifications, are sought, often reducing the chances of 'old' and traditional workers to be 'qualified enough' for the new tasks.

6.2.2   Regional and local context

The city of Santo André is one of the seven municipalities that constitute the ABC Region and has a population of 2.3 million inhabitants. Historically, the ABC region has been known as an important industrial pole located in the south-eastern part of the metropolitan Region of São Paulo

Figure 6.2   The state of São Paulo
Source: www.v-brazil.com

(MRSP). Please refer to Figure 6.2 for a map of this area. From the 1930s to 1980s, during the Brazilian import substitution phase, a greater part of industrial investments were concentrated in the metropolitan area of São Paulo. From the 1950s, the ABC region was considered an important economic and industrial area, owing to the high concentration of car manufacturing, petrochemical, plastics and metallurgy companies. Please refer to Figure 6.3.

By the 1970s, the ABC region had become Brazil's largest industrial complex, in tune with the policy of national industrialization. Subsequently, in the 1980s and 1990s, the region applied a policy of industrial dispersal and reconfiguration. Many firms that failed to restructure and boost their efficiency to confront growing international competition and globalization were forced to close. Meanwhile, industrial companies began a mass move away from the ABC area, to be gradually replaced by business

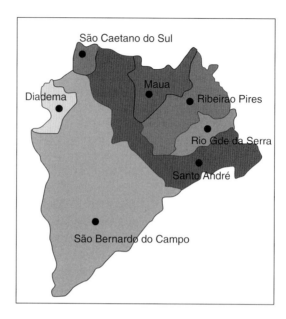

Figure 6.3 The ABC region
Source: http://www.wikipedia.org

and services companies. Santo André was hit particularly hard, with loss of value added as high as 39%. In the 20 years after 1980 the city's US$2 billion economy receded to about US$1.2 billion. These changes resulted from the federal government's decision to dismantle most of the regulatory and fiscal barriers that had previously hampered the opening of the Brazilian economy to foreign investment and capital. The problem was compounded by a shrinking municipal budget, leaving little space for the Santo André local government to take action. Other repercussions became evident, such as the emergence of more small businesses. Over the past decade, the number of large businesses (over 500 employees) in the city declined from 36 to 21, while the number of micro-businesses (employing up to 20 persons) increased from 5100 to 7600. This growth can also be observed in the commerce and services sectors. In 1989, large companies were responsible for 39% of all registered employment. Ten years later, the percentage had declined to less than 25%, while registered employment in micro-businesses increased from 15.2% to 28.5% during the same period.

An important characteristic of the ABC region is its tradition of community participation and workers' organization, especially in the industrial sectors, due to the importance of these economic activities in the area. Hence, the ABC region has also been an important base for the PT party.

In 2000, the population of Santo André was 649,331, representing 3.2% of the total population of the metropolitan region of São Paulo (and 28% of the Greater ABC region). The population of Santo André is completely urban, occupying 175 km², with a density of 3710 inhabitants per km², considered high by national standards.

The degradation of environmental and social (including labour) conditions in Santo André is reflected by the increasing number of slums. Between 1991 and 1996, the slum population increased by 3.78%, while the total population increased 0.31% (Larangeira, 2003). In 1992, there were 84 slums in Santo André. By 1997 this number had increased to 138 slums, accommodating 120,500 inhabitants (around 18% of the total households). A preliminary survey performed by the Social Inclusion and Housing Department (DEHAB) in 2001 found that: around 25,000 families, 35% of settlements and 63% of the resident population live on steep hills; 23% of all settlements and 28% of families live on floodplains or along creek and river banks; 6% of settlements and 9% of families are in environmental preservation areas; and another 16% of settlements and 18% of all families survive in areas under a range of different risk conditions (under high voltage lines, on overpasses, in waste dumps, and near oil pipelines). Only 37% of housing settlements (approximately 13% of this deprived population) are in environmentally or geo-technically safe sites' (Spertini & Denaldi, 2001: 10).

Before 1993, local authorities in Santo André had some experience in the urbanization of slums, namely in the Tamarutaca slum. Nevertheless, during the 1993–1996 administration, the deterioration of the urban situation was accentuated by the lack of an effective policy, with the focus on other priorities. In 1997, a new administration emphasized the need to make social inclusion a priority in local policies. In addition, citizen participation in decision-making and policy implementation became essential for the local government. This intention was illustrated by the creation of municipal councils. Today, there are 25 municipal councils in Santo André, each of them related to a specific matter. Every municipal council is composed of representatives of inhabitants (50%), representatives of workers (25%) and representatives of the local government (25%).

In Santo André, the majority of people living in poor conditions are concentrated in degraded urban spaces. The municipal government started an integrated policy to fight the economic and social crisis, which has generated terrible living conditions, informal and child labour, different forms of urban violence and an increasing number of people living in the streets, as well other types of social exclusion. Thus, in 2003, the Municipality of Santo André signed a memorandum of intention with the ILO for the promotion of decent work, expressing their willingness to encourage the different components of decent work in their local policies.

6.2.3 *The informal sector*

According to IBGE (2000), in Brazil just 20.2% of workers in the construction sector are registered and therefore considered as part of the formal

Table 6.6   Persons working in the formal and informal sectors in Santo André (2000)

Job status	Classification	Number of workers
Domestic worker with signed work document	Formal	4066
Domestic worker without signed work document	Informal	9601
Employee with signed work document	Formal	133,819
Employee without signed work document	Informal	39,758
Employer contributing to social security system	Formal	7010
Employer not contributing to social security system	Informal	1700
Self-employed worker contributing to social security	Formal	18,240
Self-employed not contributing to social security	Informal	31,080
Military personnel and public sector employees	Formal	9676

Source: IBGE (2000) *Pesquisa Mensual de Emprego: Janeiro de 2000.*

sector, as shown in Table 6.6. All other workers in the construction sector work without any type of legal registration. About 60% of workers in the construction sector in Brazil are not protected by the labour legislation or by any type of social coverage, such as health insurance, unemployment benefits, accident insurance coverage or pension schemes.

Informal occupations exist in smaller-scale specialized firms, which are subcontracted by the larger-scale 'parent' firms to perform specific tasks in construction. In fact, 'one of the main reasons for subcontracting small-scale firms to carry out specific tasks rather than hiring more direct labour is precisely because it is difficult for the large-scale parent firms to hire informal workers directly. Thus, through the subcontracted firms, they hire informal workers indirectly' (Werna, 1997: 20).

Another characteristic of Brazilian workers in the construction sector is the mobility between the formal and informal sectors. Werna (1997: 21) notes that 'after spending a number of years in the formal construction industry – and therefore acquiring some skills – many workers eventually leave for the informal sector'. Thus, the reasons to move to the informal sector are: better earnings and/or periods of unemployment in the formal sector. The principal motives to stay or come back to the formal sector are: 'social security and/or incapability to survive in the informal sector'. Therefore, 'a common practice is to move cyclically between formal and informal jobs'. For example, 'to stay in formal employment for the minimum period of time necessary to have access to social benefits, then resign and move to informal activities for the maximum period of time that a worker can stay outside formal employment without losing social benefits, and then to start the cycle again' (Werna, 1997: 21).

In the year 2000 in Santo André, more than 32% of workers were part of the informal sector and were present in all branches of economic activity in the municipality. As shown in Table 6.6, more than 80,000 people were employed in an informal occupation.

Table 6.7   People in work in Santo André by type of activity (2000)

Type of activity	Informal (percentage)	Formal (percentage)
Domestic services	70	30
Construction	62	38
Other collective services	53	47
Accommodation, food/drink services	46	54
Agriculture	40	60

Source: IBGE (2000) Pesquisa Mensual de Emprego: Janeiro de 2000.

This proportion is higher in the construction sector. According to the IBGE (2000), more than 60% of workers in the construction sector in Santo André had informal work status, as shown in Table 6.7.

A further indication of the greater degree of social vulnerability faced by workers in the informal sector is that they outnumber those in the formal sector living in slums. According to representatives from the municipality of Santo André, around 20% of the workers in the informal sector live in slums, while only 8.7% of formal workers do so. Furthermore, 11.9% of self-employed workers live in slum settlements, compared with 2.9% for formal sector workers.

6.3   Decent work indicators

6.3.1   Indicators of employment

Unemployment

Unemployment rates in Brazil have been estimated according to different methodologies by different institutions throughout the years. Unemployment rates are understood according to the definition proposed by the book research team: 'the proportion of the working age population unable to find work in the month prior to data collection'. At the national level, between 1992 and 2001 unemployment increased from 3.4% to 13.3%, while in construction it increased from 6.4 to 7.1%. At the local level it is difficult to study unemployment trends because information only exists for the metropolitan region of São Paulo (MRSP) for 1992, when the figure was 10.4%. However, and according to the Secretary of Development and Regional Action of the Municipality of Santo André, the unemployment rate in Santo André in the year 2001 was as high as 16.3 % of the working age population. In the year 2001 in the MRSP as a whole, the unemployment rate for the construction sector was 6.1% of the working age population. See Table 6.8.

Low wages

The average wage in the construction sector is lower than the national average in the Brazilian economy, as noted by DIESSE (2001) and based

Table 6.8   Unemployment rates

Unemployment rate		1992	2001
National level	All sectors	3.4%	13.3%
	Construction sector	6.4%	7.1%
Local level	All sectors	10.4%[12]	16.3%[13]

Sources at the national level: IBGE (1992) *Pesquisa Nacional por Amostra de Domicílios*; IBGE (2001a) *Pesquisa Nacional por Amostra de Domicílios*.
Source at the local level: IBGE (1992) *Pesquisa Nacional por Amostra de Domicílios*.

on the databases of Ministry of Labour and Employment – RAIS[3] and CAGED[4]. When measuring the remuneration of formal workers in the construction sector, in terms of minimum wages, the distribution between 1992 and 2001 shows a significant difference. In 1992, around 50% of employees in the construction sector received less than three minimum wages; while in 2001, the same amount was received by 65% of workers. Also according to DIESSE (2004: 18–20), in 2004 from 100,000 jobs generated in the construction sector, around 54% of workers received between one and one and a half minimum wages. Thus, around 30% received between one and a half and two minimum wages.

According to the Secretary of Development and Regional Action of the Municipality of Santo André, at the time of the fieldwork the Brazilian minimum wage was the equivalent to US$150 (350 reals).

As shown in Table 6.9, the percentage of employees earning less than half the minimum wage decreased between 1992 and 2001 from 13% to 7.8% at the national level. The same tendency occurred in the construction sector, with a decrease from 7% to 4.1% of employees earning less than half the minimum wage. There is no such information for the municipality of Santo André. However, in 2001, 10.2% of the employed population earned less than half the minimum wage.

Hours of work

This indicator has been calculated by measuring the proportion of wage earners working more than 44 hours a week (considered as the standard

Table 6.9   Low wage rates

Low wage rate		1992	2001
National level	All sectors	13 %	7.8 %
	Construction sector	7 %	4.1 %
Local level (Santo André)	All sectors	n/a	10.2%

Sources at the national level: IBGE (1992) *Pesquisa Nacional por Amostra de Domicílios*; IBGE (2001a) *Pesquisa Nacional por Amostra de Domicílios*.
Source at the local level: Prefeitura de Santo André (2002) *Santo André: Integracion de Programas para promover la inclusion social*.
Measured by the number of employed persons earning less than half the minimum wage in 1992 and in 2001.

Table 6.10   Hours of work

Hours of work		1992	2001
National level	All sectors	39.3%	40.9%
	Construction sector	50.4%	50.3%
Local level (Santo André)	All sectors	41.9%	54%
	Construction sector	38.8%	51.7%

Sources at the national level: IBGE (1992) *Pesquisa Nacional por Amostra de Domicílios*; IBGE (2001a) *Pesquisa Nacional por Amostra de Domicílios*.
Sources at the local level: Municipality of Santo André website; IBGE website, data at the municipal level.

number of working hours per week according to the Brazilian labour legislation). Throughout the 1990s, the proportion of construction workers working more than 44 hours a week significantly increased, most notably in Santo André, from 38.8% to 51.7%. For the whole Brazilian economy, this rate stayed mostly stable, with a slight increase (from 39.3% in 1992 to 40.9% in 2001). However, in the case of Santo André, the general tendency followed that for the construction sector, as the number of workers doing more than 44 hours increased from 41.9% in 1992 to 54% in 2001.

It is important to note when analyzing this data that it often includes the informal sector or autonomous workers. Another important qualifier is that often in the construction sector work is done by periods, depending on the sites under construction at a precise moment. Hence, the proportion of working hours can sometimes be lower than indicated. See Table 6.10.

Health and safety at work

There is not much information about safety at work that can be gained from Brazilian data sources. Only in recent years has safety at work has been formally included in statistics from the database of the Ministry of Labour and Employment – RAIS and CAGED. Data specifically concerning the construction sector was not easy to find for the two selected periods. However, DIESSE presented some information concerning the number of accidents registered in the construction sector in Brazil for the period 1998–2001.

According to DIESSE (2001: 13–14), in Brazil data about work accidents comes from the Ministry of Social Security and refers to the concept defined by the 8.213/91 law and the 3.048/99 decree. The total number of registered accidents concerns the number of accidents processed by administration at the National Institute of Social Security (INSS). As shown in Table 6.11, in the construction industry, at the national level, the percentage of accidents decreased from 31,959 in 1998 to 25,627 in 2001. The number of fatal accidents also decreased from 448 in 1998 to 337 in 2001.

Table 6.11   Number of accidents registered in the construction industry
(1998–2001)

Sector	Number of work accidents registered					
					Reason	
	Year	Total	Typical	Way	Work/illness	Death
Construction	1998	31,959	29,060	1963	936	448
	1999	27,826	24,950	2008	868	407
	2000	25,536	22,637	2112	787	325
	2001	25,627	22,741	2181	705	337

Source: DIESSE (2001) Os Trabalhadores e a Restructuração Produtiva na Construção Civil Brasileira, p. 14.

6.3.2   Indicators of social security

Public social security coverage

In Brazil, social security coverage is more widespread in the southeast and southern regions (such as São Paulo), also known as the industrial regions. In the construction sector, public social security coverage is lower than in other economic sectors. Concerning the gender dimension, in general for all sectors, men have greater social security coverage than women. However, the opposite situation occurs in the construction sector. In fact, the coverage rate among women has not only increased but today is higher than for men. See Table 6.12.

The public health care system, the Sistema Unica de Saude (SUS), was established in 1988 with the aim of providing universal care. It is funded through federal and local taxation and by contributions from employers and employees. Nevertheless, the SUS system has financial problems: 75% of the population is dependent on it, but only 40% of all health spending comes from public resources. Owing to the inadequacy of services, the majority of middle and upper class Brazilians have additional medical insurance, mostly through contracts between their employers and private health care insurance companies.

Table 6.12   Public social security coverage

Public social security coverage		1992	2001
National level	All sectors	43.4%	45.7%
	Construction sector	38.7%	27.61%

Sources at the national level: IBGE (1992) Pesquisa Nacional por Amostra de Domicílios; IBGE (2001a) Pesquisa Nacional por Amostra de Domicílios.

Table 6.13   Old age pension

	1992	2001
National level	7%	13.3%

Sources: IBGE (1992) *Pesquisa Nacional por Amostra de Domicílios*; IBGE (2001a) *Pesquisa Nacional por Amostra de Domicílios*.

Old age pension

Indicators on old age pensions have not been easy to find. Data has been found at the national level only. The IBGE highlights that in 1992, only 7% of people aged 65 or older benefited from a pension scheme. In 2001, the old age pension was attributed to only 13.3% of people aged over 65 years. See Table 6.13.

6.3.3   Indicators of workers' rights

Legislation on workers rights and working conditions

Relations between workers and employers in Brazil are ruled by the Consolidation of Labour Laws[5] (CLT). The CLT is very wide ranging and detailed, regulating the most varied aspects of labour relations. The CLT was first introduced in 1943 during the administration of President Getúlio Vargas in order to consolidate the labour laws existing at that time. It was created as a system to protect workers from exploitation by employers and to harmonize labour relations avoiding direct disputes between parties. Little space was left for direct negotiations between employers and employees, as the law required disputes to be settled in labour tribunals rather than in the companies involved and discouraged the development of a cooperative relationship.

By definition, a formal sector worker has a working card (*carteira assinada*) signed by his employer. Besides the obligation to sign the card, the law stipulates a set of minimum conditions that any employment relationship must follow. The most important rules include:

- Minimum wage
- Maximum hours of work per week
- Minimum payment for extra-time work
- Maximum extra-time working hours
- Pre-paid annual vacations
- Special protection clauses for women and children
- The dismissal of pregnant women is forbidden
- The right of paid vacation before and after childbirth for the mother
- Special work conditions for nightshifts, one month pre-notification of firing, and protection against unjustified dismissals

Since the creation of the Consolidated Labour Laws (CLT), there have been changes in the legislation. In particular:

- In 1962: introduction of a one-month's wage annual bonus (thirteenth salary).
- In 1963: introduction of a family allowance.
- In 1965: introduction of a wage adjustment law which determines the minimum rate of wage adjustments of all workers in the economy.
- In 1966: creation of a severance fund[6] in place of a clause forbidding dismissal of workers with more than ten years' tenure.
- In 1986: creation of an unemployment insurance program, which today covers about 25% of the country's labour force.
- In 1988: approval of a new constitution with the introduction of new labour clauses.

The main changes to labour legislation introduced by the constitution of 1988 were:

- The maximum number of hours of work per week changed from 48 to 44 hours and the minimum payment for extra-time hours increased from 20% to 50% of the worker's wages.
- Maternity leave for mothers was increased to 120 days and five days childbirth leave for the father was introduced.
- A vacation bonus of one-third of the workers' wages was created.
- The number of daily working hours was reduced from eight to six hours for continuous work shifts.
- Firing costs for unjustified dismissals increased from 10% of the FGTS balance to 40%.

The Ministry of Work and Employment also specified correct conduct through regulatory standards[7]. For instance, one of the most important regulatory standards in the Construction sector was the NR-15, which states the basic rules for unhealthy operations and activities.

Another important component is the Brazilian Unemployment Insurance (UI). This insurance system is characterized by a low replacement ratio, a short benefit duration and the fact that it is restricted to workers in the formal sector. This implies that UI is not accessible to over half the workforce. Although present in the constitution since 1946, it was not until 1990 that UI became universally accessible. The current Brazilian unemployment system was created in 1986 within the context of the Cruzado Plan. In 1988, the source of funding was changed from general treasury revenues to the Fundo do Amparo ao Trabalhador (FAT)[8]. Eligibility criteria were relaxed in 1990 (Law No. 7.998), expanding the base of workers with access to UI benefits. By 1990, UI covered 43% of all dismissals from

Table 6.14   Wage inequalities between genders

Wage inequalities between genders		1992	2001
National level	All sectors	1.88	1.59
Regional level*	All sectors	1.86	0.84
	Construction sector	0.88	0.91

Sources at national level: IBGE (1992) Pesquisa Nacional por Amostra de Domicílios; IBGE (2001b) Pesquisa Nacional por Amostra de Domicílios.
Sources at the regional level: IBGE (1992) Pesquisa Nacional por Amostra de Domicílios; IBGE (2001a) Pesquisa Mensal de Emprego: Janeiro 2001.
* Presented for the metropolitan region of São Paulo, and not for Santo André.

formal employment. UI benefits in Brazil are low and do not exceed two minimum wages.

An important feature of Brazil's labour legislation is the coexistence of individual and collective employment contracts. Individual contracts are concluded between the company and the worker and deal with issues such as working conditions and wages. In contrast, collective contracts are concluded between the employer and the workers union, or between the employers' association and the union. These contracts cover minimum working conditions and minimum wages, among other issues.

Wage inequalities between genders

Indicators in Table 6.14 show that in all sectors at the national level, average wages are more than 50% higher for men than for women. Nevertheless, over the years women have occupied more job places, transforming this indicator in the case of São Paulo, where the situation had completely changed by 2001. In the construction sector, indicators show a better situation for women, who occupy more skilled jobs than men. This is also due to the fact that in the construction sector, working women occupied more experienced and concrete positions than men (such as engineers, architects or services to the construction workers, such as delivery of food, etc.).

Wage inequalities according to workers' places of birth

It is important to point out that none of the institutions consulted have statistics concerning wage inequality between natives and foreigners. This analysis according to nationality is not commonly used in Brazilian statistics and other sources of data.

Child labour

In Brazil, according to the International Program on the Elimination of Child Labour (IPEC) from the ILO (2003d: 14–16), more than eight million children between five to seventeen years old did some kind of work

Table 6.15   Quantity of young people (5–17) per type of work

Year	Domestic workers	Percentage	Non-domestic workers	Percentage	Non work	Percentage	Total
1992	882,807	1.9	7,540,641	17.6	34,643,607	80.5	43,067,055
1999	502,839	1.2	5,989,906	14	36,303,317	84.8	42,796,062

Source: ILO (2003d) (IPEC) Brazil Child and Adolescent Domestic Work in Selected Years from 1992 to 1999: A National Report, p. 14.

(with or without remuneration), representing 19.5% of all children of that age group in 1992. Data shows that 17.6% (7,540,641) were occupied in a non-domestic activity and 1.9% (882,807) in a domestic activity. This proportion was reduced in 1999/2000, when 15.2% of children aged five to seventeen were working; 14% (5,989,906) in a non-domestic activity and 1.2% (502,839) in a domestic activity. As noted by the IPEC, the ratio of children who did not work increased on average 0.7% per year during the same period. See Table 6.15.

In Brazil, the official minimum age to be employed is 16, but the Child and Adolescent Statute accepts children over the age of 14, as long as the child is an apprentice. According to Education International (2004: 56), child labour constitutes a big problem in Brazil. Legally, people under 18 should not work in dangerous activities. However, thousands of children work in conditions prohibited by the ILO Convention 182 on the Worst Forms of Child Labour. According to the Brazilian Government and as noted by Education International (2004: 56), in 2000 around 60,000 children between seven and seventeen years of age were working in bad conditions, namely in the rural areas such as in cane plantations. Another significant problem is child prostitution, which is widespread in some regions of the country. Policies to eradicate child labour, which have been enforced by the federal, state and municipal authorities; as well as specific legislation that forbids child labour, have contributed to declining trends.

According to the IPEC (ILO, 2003d: 23), the majority of children working are girls. Between 1992–1999, out of 100 children (five to seventeen) involved in domestic labour activities, approximately ninety-five were girls and only five were boys. This national proportion is the same in the regional areas. Another important element concerning the demographic characteristics of working children is that, for those engaged in non-domestic activities, the relation is inversed and boys are predominant. Thus, in the period 1992–1999, 73% of children involved in non-domestic activities were boys. In this case, the proportion is higher in the north (78%) and centre-west (77%) regions. The proportion in other regions is less than the national average: in the northeast (72%), southeast (72%) and south (67%). See Table 6.16.

Santo André is the second largest city in the metropolitan region of São Paulo, with a total population of 649,331 inhabitants in 2000, and 9.5% of

Table 6.16 Child labour

Child labour*		1992	1999/2000
National level	All sectors	19.5%	15.2%
Local level (Santo André)	All sectors	n/a	1.5%

Sources at the national level: ILO (2003d) *Brazil Child and Adolescent Domestic Work in Selected Years from 1992 to 1999: A National Report*; IBGE (1992) *Pesquisa Nacional por Amostra de Domicílios.*
Source at the local level: Santo André Municipality Department of Workers' Education.
* In this case, children aged five to seventeen years old are taken into consideration.

them aged between 15 and 19 years. According to the Santo André Municipality Department of Workers' Education (Prefeitura de Santo André, 2004c: 2) and the IBGE (2001a), unemployment among youth between 16 and 24 years old represented 17.8% of the total unemployment in the region.

There is no specific data concerning child labour in Santo André for the year 1992. However, according to the Municipality of Santo André, 836 of the 55,090 children aged between ten to fourteen years were working in 2000. This corresponds to 1.5% of the total number of children in that age group. Most working children were in the urban areas and none in the construction sector.

6.3.4 *Indicators of social dialogue*

Workers without an individual contract are not covered by collective contracts in their job category. In Brazil, the results of collective bargaining processes are legally extended to all workers and companies in that specific sector or industry, even if the workers or companies involved are not members of the particular unions negotiating the agreement. This legal stipulation gives much importance to workers' unions in labour relations.

Legislation on social dialogue

In the legal arrangements originally established by the consolidated labour laws (CLT), unions were responsible for contributing to the harmonization of relations between capital and labour, and helping to implement the Government's economic policies.

The principles on which labour legislation and organization were founded gave labour unions power and close links with the state, which compromised their action in support of workers' interests. Under the CLT, labour unions are organized by occupational category, but employers' associations are organized by economic sector. Job categories and economic sectors are defined by the Ministry of Employment and Labour on the basis of similar characteristics. Until 1988, different occupations and economic categories were prohibited from grouping together in a single union. This restriction was lifted in the 1988 Constitution, when the formation of nationwide unions and union confederations was authorized.

The smallest regional base is the municipality, but unions can also have regional, state, or even national jurisdiction. All collective bargaining processes in a given category have to be carried out with participation from the union holding the monopoly representation in the geographic area concerned. Although union membership is not compulsory, workers and employers are required to pay a union tax annually. The Ministry of Employment and Labour passes 60% of this on to the respective union, and is also the body responsible for collecting it. The remaining funds are divided between the Ministry of Employment and Labour and the federation and/or confederation for the occupational or economic category concerned. By law, funds transferred to labour unions must be used exclusively for purposes such as recreation, social assistance, education or cooperatives, but never to finance political activities, collective bargaining processes or strikes. Only funds obtained through voluntary contributions can be used for such purposes.

Until 1988, the Ministry of Employment and Labour could interfere in labour unions' activities for reasons such as misuse of the union tax or calling unauthorized strikes or lock-outs. The law even enabled authorities to abolish a union if it was judged to have impeded the implementation of Government economic policy. The Constitution of 1988 eliminated this provision.

Collective bargaining is compulsory and must take place once a year, during the 'base-date' period, in which the worker's union and the employer's organization or individual company negotiate wages and other employment issues. Base-dates vary between occupations and categories. However, for different occupations it is possible to sign agreements on the same day in the same company or economic category, thereby resulting in a collective agreement covering a large proportion of the workers in a given industry. In bargaining processes between employers' organizations and workers' unions, the result is known as a collective convention (*convenção coletiva*). If bargaining takes place between the workers' union and a single company, the outcome is known as a collective agreement (*acordo coletivo*). Until 1988, any contract or agreement contrary to the Government's overall economic or wage policy was susceptible to legal annulations. Since then, unions have been free to negotiate agreements without the threat of Government interference.

The new constitution in 1988 was the basis for modernized legislation on collective bargaining, wages, strikes and unions. Collective bargaining began to be encouraged, the concept of 'illegal strikes' ceased to exist, prohibition of worker organization at the national level was lifted and the participation of civil servants in unions was allowed, along with other changes which made union organization more democratic. Nevertheless, these changes were insufficient to reduce the greatest obstacles in

Table 6.17   Union density rates

Union density rate		1992	2001
National level	All sectors	20.2%	20%
	Construction sector	10.2%	7%

Sources: IBGE (1992) *Pesquisa Nacional por Amostra de Domicílios*; IBGE (2001a) *Pesquisa Nacional por Amostra de Domicílios.*

implementing collective bargaining processes in Brazil, since the basic rules governing employment contracts, the powers of labour tribunals and union organization remained virtually intact.

Union density rate

Table 6.17 shows the proportion of unionized workers in the total employed population. Available data indicates that union density fell from 20.2% to 20% between 1992 and 2001 at the national level. Even if not significant, this decline may be associated with changes that have taken place in the economy and the labour market over the last two decades, particularly in the 1990s, when trade liberalization policies were introduced. Other potential explanatory factors include market deregulation, including the labour market, and the privatization of state-owned companies (where unionization was traditionally active). These changes increased competition, undermined profits and forced companies to adopt cost-cutting measures, potentially affecting workers through a decrease in the number of available jobs. A drop in union density was an expected consequence of this situation, resulting from higher unemployment and less chance of success in wage claims.

It was not possible to find any information on union density for the municipal level of Santo André or São Paulo.

Collective bargaining rate

According to the ILO office in Brazil, this rate cannot be calculated in Brazil owing to the organization of trade unions by professional category combined with geographical location. Thus, collective bargaining comprises all the workers represented by the union, the workers' federation or confederation, irrespective of whether they are active members of a union or not.

6.3.5   *Synthesis: Santo André decent work indicators*

Table 6.18 presents the decent work indicators for Brazil and Santo André according to the four key components. The right hand column of this table

Table 6.18   Santo André decent work indicators

Employment dimension				
Unemployment rate				
		1992	2001	Trend towards DW
National level	All sectors	3.4%	13.3%	−ve
	Construction	6.4%	7.1%	−ve
Local level	All sectors	10.4%	16.3%	−ve
		(MRSP)	(Santo André)	
	Construction	n/a	6.1%	
Low wage rate				
		1992	2001	Trend towards DW
National level	All sectors	13%	7.8%	+ve
	Construction	7%	4.1%	+ve
Local level (Santo André)	All sectors	n/a	10.2%	
	Construction	n/a	n/a	
Hours of work				
		1992	2001	Trend towards DW
National level	All sectors	39.3%	40.9%	−ve
	Construction	50.4%	50.3%	
Local level (Santo André)	All sectors	41.9%	54%	−ve
	Construction	38.8%	51.7%	−ve
Social security dimension				
Public social security coverage				
		1992	2001	Trend towards DW
National level	All sectors	43.4%	45.7%	+ve
	Construction	38.7%	27.61%	−ve
Old age pension				
		1992	2001	Trend towards DW
National level	All sectors	7%	13.3%	+ve
Workers' rights dimension				
Wage inequality between genders				
		1992	2001	Trend towards DW
National level	All sectors	188%	159%	−ve
	Construction	n/a	n/a	
Local level (MRSP)	All sectors	186%	84%	−ve
	Construction	88%	91%	−ve
Child labour				
		1992	2001	Trend towards DW
National level	All sectors	19.5%	15.2%	+ve
Local level	All sectors	n/a	1.5%	
Social dialogue dimension				
Union density rate				
		1992	2001	Trends towards DW
National level	All sectors	20.2%	20%	=ve
	Construction	10.2%	7%	−ve

shows trends towards (positive) or away from (negative) decent work for each of the indicators in all sectors and in the construction sector at the national, regional (MRSP) and municipal (Santo André) levels.

The indicators presented in Table 6.18 show both positive trends towards and negative trends away from decent work in Santo André and Brazil during the 1990s. It is evident that the indicators for the employment dimension and the workers' rights dimension are the most negative. These negative trends ought to be attributed high priority at both the national and local levels in order to promote decent work in Brazil.

As noted above, much data is not available for the municipality of Santo André. Therefore, in some cases Table 6.18 had to be based on local data for the metropolitan region of São Paulo (MRSP) and the ABC region (which includes Santo André as well as other municipalities). If these or other local authorities wish to monitor trends towards or away from decent work then new initiatives will be necessary in order to collect and analyze information at the municipal level.

6.4   Decent work in Santo André: best practices

6.4.1   The 'Santo André Mais Igual' (SAMI) program

Presentation and context

Since the early 1990s, the municipality of Santo André has been known for tackling social exclusion. After different trials, a strategy was drawn up to embrace a new method for driving city management and local public policies, resulting in the Santo André Mais Igual[9] program (SAMI) launched in 1998. Initially known as the Integrated Program of Social Inclusion, the SAMI program aims to reduce inequality in the municipal area and simultaneously deal with the many aspects of social exclusion by incorporating community participation. It has been found that direct community participation is vital: identifying needs, discussing what is being done and what needs to be done. Communities directly participate in:

(1) The definition of the territory that will benefit from the participative budget program (Presupuesto Participativo)
(2) The permanent follow-up of the urban, economic and social proposal interventions
(3) The local team composition, namely the community health agents, credit agents, educators, instructors and others

Santo André is located in the Greater ABC Region, at the south-east side of the metropolitan region of São Paulo, and in 2000 it had 649,331 inhabitants. Of the total Santo André population, 10.34% lived in a poverty situation and 5.06% in a misery situation (Denaldi, 2004)[10]. Thus, 15.43% of the Santo André population lived with a monthly income per capita equal to or less than half the minimum salary.

The situation before the implementation of the SAMI program was very delicate. According to the Municipality of Santo André (2002) and as stressed by Denaldi (2004: 2), in 1998 about 20% of the Santo André population lived in 138 slum nuclei. Among the heads of families, 51.4% were unemployed and/or underemployed, 13% were illiterate, and 25% had not advanced beyond primary school. In terms of families, there were 57.6% female households. Single-parent families headed by women constituted 32.7% of those families with earnings of less than half the minimum wage. Furthermore, children and adolescents were openly exposed to crime and the lack of basic sanitary services was a serious problem.

Four high-risk slum nuclei were selected for the first phase of the SAMI program: Tamarutaca, Sacadura Cabral, Capuava and Quilombo II, each of them with its own concerns (Larangeira, 2003). Tamarutaca and Capuava were subject to landslides, both of them being built on steep slopes and Sacadura Cabral endured flooding. The Participatory Budget Council made the urban renewal of Capuava and Tamarutaca their top priority in 1997, while Sacadura Cabral and Quilombo II were targeted for renewal in 1998. Land ownership was the key to the interventions planned for these areas. See Figure 6.4.

Depending on the social priorities in each locality, site selection for implementation of the SAMI program was determined by two major criteria: technical and political. Policy-related criteria included priority-care issues as set out in the Municipality Master Plan (itself based upon the Participatory Budget), free discussion with local leaders (where the Santo André Movement for Protection of the Slums Residents' Rights has played a key role) and community empowerment and organization. Technical issues included the impact of environmental rehabilitation on the quality of life, the legal status of land mainly occupied by slums and the financial resources available from each partner. Property tenure in each slum as well as individual interventions were additional considerations for selection, with emphasis on the longer-established settlements.

Given that slum upgrading that focuses only on physical and environmental aspects does not significantly alter the social exclusion of residents, this program aimed at bringing together actions of various sectors and departments from the municipality of Santo André in order to jointly identify where social exclusion was severe, as well as to create the conditions required to improve the social and economic well-being and quality of life of the target populations. Eighteen different programs exist and are in progress. They involve the combined efforts of 12 municipal departments coordinated by the Social Inclusion and Housing Department, which is responsible for managing the SAMI program. This strategy made it possible for the municipality to take a range of complementary actions forward. At the same time, the structure of the municipal administration was modified

Figure 6.4   Poverty location in Santo André
Source: Municipality of Santo André.

in order to introduce a more interactive and flexible matrix model with improved communication channels. This new model provided an organizational structure that gave priority to the participation of all the actors involved, from decision-making to action in the field.

By acting on all the different facets of social exclusion, the SAMI program seeks to ensure interaction between the various sectoral programs and to focus on the main problem areas. As highlighted by Celia Chaer, coordinator of social inclusion in the Santo André municipality, the SAMI program is unique because the efforts of the different departments that collaborate on different initiatives share the same objective. This is synthesized in the motto: 'All together, at the same time, in the same place'.

The SAMI program comprises three dimensions:

(1)   The urban dimension: the main objective is to integrate the slums into urban infrastructure by ensuring access to urban equipment, community services and sanitary housing. The SAMI program is meant to improve slums and to initiate improvements in housing conditions by providing urban infrastructure, guaranteed land tenure and assistance for self-help construction projects.

(2)   The economic dimension: the objective is to improve income and employment conditions through different programs including Minimum Income Guarantee, Social Interest Job Generation, Professional Education, Cooperative Incubators, Popular Entrepreneurs, People's Bank and the Centre of Autonomous Services. Besides all these, the Business and Services Office brings together economic units managed by enterprising neighbours of the slum nuclei, such as the Centre of Autonomous Services.

(3)   The social dimension: this focuses on activities and programs related to education, health, social assistance and culture. Some of the programs are the MOVA/SEJA Program, the Family Health Program, the Child Citizen Program, the Little Seed Program and the Rights Counter Program.

A self-help building project in slums was a significant component of the SAMI program. Among the first four slums selected for the SAMI program, Sacadura Cabral stands out from the others owing to its achievement in employment generation and promoting decent work, and its successful efforts in the construction sector. The area known as Sacadura Cabral is situated in the north-west of the Santo André municipality in the suburb of the same name, bordering the city of São Bernardo do Campo. It is a consolidated slum settlement despite recent signs of outward growth. Urbanization in Sacadura Cabral began with a small association of residents who were trained in construction activities. This then expanded to the rest of the slum residents who wanted to build their own houses. The implementation of self-help building was extended beyond individual houses to collective assistance for neighbours. Some residents who built their own houses in Sacadura Cabral are now working in the construction sector. Between 1999 and 2003, 770 families participated in the auto-construction project in Sacadura Cabral. The fact that most houses have connections to public services (water, sanitation services, electricity, etc.) has increased residents' quality of life of as well as their self-esteem. In addition, house prices have increased significantly. Other repercussions of this project include a reduction in vandalism and an increase in the self-confidence of residents.

Table 6.19 presents an overall picture of beneficiaries in the nine nuclei slums where the SAMI program was implemented.

In the following points we will briefly present some projects from the SAMI program, particularly those related to employment generation and decent work in the construction sector.

Let's Build

Let's Build (*Vamos Construir*) is a professional training project for adults in the construction sector that ran between 2002 and 2003. It was managed by

Table 6.19 SAMI Program phases and beneficiary slums

Slum name	Surface (m²)	Housing units
Tamarutaca	109,705	1269
Sacadura Cabral	42,259	549
Quilombo II	12,739	230
Capuava	98,404	1327
Espiritu Santo I	161,648	597
Gonçalo Zarco	2680	148
Marginal Guarará	17,500	63
Capuava Unida	10,920	295
Gamboa I	40,587	560

Source: Prefeitura de Santo André (2002) *Santo André: Integracion de Programas para Promover le Inclusion Social.*

the Department of Education and Professional Training of the Municipality of Santo André. The Let's Build program was part of the urbanization program managed by the Secretariat of Social Inclusion and Housing.

In October 2002, a hundred families benefited from this project, which focused on urban upgrading in SAMI selected areas. Its main goal was to provide professional training in construction for beneficiary families, enabling them to build or improve their own home. After constructing their homes, those people who have been trained are in a better position to find a job in the construction sector and secure decent work and a better quality of life. Some participants in the Let's Build scheme in Sacadura Cabral became professional construction workers.

Centre for Autonomous Services (Central de Serviços Autonomos)

The experience of self-help building has shown the excellent construction skills of some of the participants in Sacadura Cabral. The Centre of Autonomous Services began operations in October 2003 in the Sacadura Cabral Business Centre. The Centre of Autonomous Services is an employment and income-generation program run by the Department of Employment and Income Generation, which is part of the Department of Development and Regional Action for the Municipality of Santo André. It is implemented by Coop Mutual Action, which is a multi-professional team composed of educators/trainers, economists, psychologists, social assistants and lawyers. Professionals from the Centre for Autonomous Services received integrated training covering diverse technical aspects in an initial training program, as well as through the Worker Education Process.

The main objective of the Centre for Autonomous Services is to train and advise on the provision of autonomous professional services in construction, renovation, painting, electricity, plumbing and finishing. This initiative has increased employment opportunities for workers in the construction sector. Workers in the construction sector were chosen to form the first database as a pilot project. Anyone who offers construction services can have his or her name on the database. The service is cost-free, with only one condition: beneficiaries must attend monthly meetings where training programs are offered.

The centre offers the following services: psychological follow-up, contracts to ensure worker's rights, a register of all contacts, advertisements, handouts, uniforms and identification visit cards. When jobs are completed, the office contacts a number of clients in order to ascertain whether they are satisfied with the workers and the quality of the job done. This is followed up with a brief report handed to the worker. In 2006, 130 professionals were enrolled: 126 men and 4 women.

Since the centre was opened, over 1300 calls have been received; a substantial number considering the size of the program and the short time that it has been in operation. Following its success, four more offices will be opened in other settlements, with a central office located near the Construction Workers Union in Santo André. To date, 1400 clients have attended and about 20,000 workers are registered each year.

Achievements and results

The first phase of the program (1997–2000) benefited about 3740 families (representing 16% of population) in the four selected slums: Sacadura Cabral (700 families), Tamarutaca (1400 families), Capuava (1400 families) and Quilombo (240 families). During the second phase of the program, between 2001 and 2004, about 2200 families benefited from the SAMI program, namely in the slums of Mauricio de Madeiros, Espíritu Santo I and Gonzalo Zarco. To date, the program has benefited 20% of the slum population in Santo André.

The program has produced positive results in the quality of urban life and community cohesion. For example, the expansion of new formal and informal businesses in the region reflects the program's policy of employment generation and income development.

Health indicators also demonstrate substantial improvement resulting from the work done by health agents. Community participation shows the greater effectiveness and consolidation of community organization. Finally, the increased enrolment of adolescents and adults in literacy and professional training courses indicates that residents have begun to understand the importance of these courses for their future.

Denaldi (2004: 12) states that an evaluation covering the period 1998–2000 shows some of the important achievements of the SAMI program. Out of the 15,000 inhabitants in the four selected slums, 2500 were trained professionally in the outline of the Citizen Worker program. No less than 16% declared that they found a job after this training. Approximately 27% of the total number of families received financial benefits from the Minimum Income program.

The SAMI program has been recognized at both the international and national levels. In 2000, the SAMI program won the award for Public Management and Citizenship[11] from the Getúlio Vargas and Ford Foundations. It was listed as one of the five best experiences of public policies developed in Brazil. In 2001, the SAMI program was included in the 16 best practices in the world, chosen to be presented at the United Nations Conference on Human Settlements Istanbul +5. Also in 2001, the SAMI program was selected as one of the ten best practices – with special emphasis on the urbanization of the Sacadura Cabral slum – and it won the Federal Caixa Economica Award for Best Practices in Local Management. In 2002, the experience Gender and Citizenship, carried out within the framework of the program, was selected as one of the ten best initiatives in the world, winning the Dubai International Best Practices Award for the United Nations Human Settlements Program (UN-HABITAT).

The municipality of Santo André has also undertaken some other policies and programs. The following sections of this chapter present those initiatives specifically related to employment generation and the promotion of decent work in the construction sector and related services.

6.4.2   The Public Centre for Employment, Labour and Income (CPETR)

The Public Centre for Employment, Labour and Income (CPETR) represents the public system of employment in Santo André. Its main objectives are to unify actions developed by the Santo André Income and Employment Centre (CTR) and the Centre of Solidarity to Workers (CST) to integrate municipal programs of employment and income generation, and activate and extend the range of services offered to employers and workers.

Client enterprises have many advantages in using the services of the CPETR, such as elimination of costs and increase of efficiency in the process of worker selection, a qualified team for monitoring and infrastructure availability. About half a million workers have been recorded since 1999; young adults entering the labour market, practitioners, technicians, specialists and professionals with handicaps. From May to August 2006, about 600 persons were placed each month.

In terms of the construction sector specifically, the CPETR has registered 33,928 workers. The main activities are: general workers (32,667), technical workers (547), and engineers and architects (714).

In Santo André, there are two local offices of the CPETR, where a thousand workers arrive per day. According to Noe Cazetta and Tiago Nogueira, from the Employment Generation Secretariat (Municipality of Santo André), there is a project for constructing the CPETR database at a regional level for the Greater ABC region.

6.4.3   Integrated Program for Qualification, PIQ (Programa Integrado de Qualificação)

The Integrated Program for Qualification (PIQ) was conceived to support young people (18–26 years old) in a risky social situation. The main idea was to establish a combination of high education (middle school level) and professional training, realized over a period of three years. The PIQ program was implemented in Santo André at the beginning of 2003 by the Department of Workers' Education. It is regulated by municipal law, following the directives of the local government: inclusive education and integral training for individuals.

The PIQ program offers courses in different areas, such as construction, mechanics and communication. It has an agreement with the Department of Employment Generation and Income, which informs trainers and trainees about important questions related to employment generation. The idea is to make a Solidarity Sector Network viable. It will include different programs in order to promote the integration of young people into the labour world.

This program helps young people from the Santo André slums. According to the Education and Professional Training Secretariat (Municipality of Santo André), the majority of the population living in risky areas demonstrate little or no school attendance and have a low income. A great number are unemployed or work under precarious conditions without any formal registration or social protection. The PIQ program includes different public centres for professional training, each of them in a specific location. Concerning the construction sector more specifically, the Armando Masso centre has provided qualifications for 420 young people in 13 different types of courses in 2006.

6.4.4   Selective Collection and Income Generation Program

The Municipal Environment and Water Company (SEMASA) is involved in a series of actions that contribute to opening up job opportunities through training programs and income creation for excluded residents. In Santo André, the Selective Collection started in 1997 with a pilot project undertaken in a number of neighbourhoods and public spaces. The objectives were to reduce pressure on the garbage dump, improve the quality of life of the resident population and generate jobs and income through the retailing of recyclable materials. Since 2000, selective door-to-door collection has been pursued throughout the city. The SEMASA Program has recorded a participation rate of 60% of the population according to a recent survey conducted by the company. This rate can be considered high even by industrialized countries. With five different programs, the SEMASA Program benefits about 282 individuals and their respective families.

The programs benefited are the following:

(1) *Coopcicla*: made up of previously unemployed people who now work screening and selling recyclable materials. The income of each member varies in accordance with the monthly sales of the cooperative but is normally in the region of US$170.

(2) *Cooperativa Cidade Limpa* (Clean City Cooperative): people work screening waste materials and in community groups collecting waste in handcarts from places inaccessible to the conventional collection vehicles.

(3) *Usina de Triagem e Reciclagem de Papel* (Selection Plant and Paper Recycling): at present this comprises 35 socially vulnerable young people aged between 14 and 17. In the paper screening and recycling plant these workers learn to recycle paper and produce handcrafted articles that are sold for pocket money. This project is coordinated by the municipality and SEMASA in partnership with the NGO Reciprocity Plant (*Usina da Reciprocidade*) which has received awards both in Brazil and from abroad.

(4) *Estação Bosque* (Woods Station): this brings together former garbage scavengers with the aim of offering better working conditions and quality of life to the participants by providing a suitable space for taking in donations of recyclable materials, thereby eliminating the need for the individuals concerned to work the streets collecting garbage with handcarts.

(5) *Projeto Refazer* (Remake Project): the patients of the Mental Health Program run by the Health Secretariat of Santo André are the beneficiaries of this program. They produce handicrafts and undertake manual work, which helps to improve their health as well as their quality of life. Materials for this project come from the homes of families that support the project.

These projects certainly contribute to social inclusion. Around 500 tons of recyclable materials collected every month are delivered to cooperatives and other social programs that display the materials before selling them to recycling plants.

6.5 Decent work: synthesis and recommendations

The compilation of decent work indicators has clearly shown both positive trends towards and negative trends away from decent work in Santo André, the ABC region and Brazil.

The absence of reliable data at the local level highlights the need for new research initiatives, at both the local and national levels, to improve current shortcomings. Unless the 'missing information syndrome' is addressed by reforms in data collection and processing it will not be possible to define and monitor trends towards or away from decent work.

This case study has included interesting projects and programs specifically concerning the construction sector and related services. Since 1998, the SAMI program has been an important initiative of the municipality of Santo André to reduce social inequalities, deal with different aspects of social inclusion and generate employment for different categories of residents. The SAMI program has been recognized at international and national levels as an example of good practice in social inclusion policies and urban public management. In essence, municipal authorities in Santo André have succeeded in guaranteeing social dialogue, one of the four pillars of decent work. The other components of decent work: employment conditions, social security and workers' rights, are also being promoted and developed by the different programs and projects implemented by the municipal authorities.

Although the achievements of the SAMI program are evident, some criticism can be made. Given that one of the principal concerns of the SAMI program is employment generation – which has mainly occurred in the informal labour market – there is no concern about formalizing these jobs or guaranteeing social protection of the workers involved. Even if workers engaged in these projects are autonomous and benefit from local registration, they are not formal workers. This can be considered a contradiction because the local authority enables informal work by autonomous workers without addressing social security. Bringing informal employment into the formal sector is a major challenge, not only for Santo André but for Brazil as a whole.

This chapter has shown how the concept of decent work in the construction sector can be promoted by a local authority. The example of Santo André can contribute to discussion on the genuine possibilities of applying this concept at the local level. However, much still remains to be done to address the negative indicators identified in this research.

This Brazilian case study is a significant example of the promotion of decent work. In 2003, the municipality of Santo André confirmed its willingness by signing a memorandum of intention for the promotion of decent work. In addition, the national government made a commitment to the cause when the President of Brazil signed a Special Agreement Program of Technical Cooperation for the Promotion of a Decent Work National Agenda in June 2006.

After the completion of field research in Santo André in October 2006, the local government began to implement new initiatives to promote decent work in conjunction with the ILO, the Canadian International Development Agency (CIDA) and the University of British Columbia. The municipality of Santo André is committed to a new project concerning the employment conditions of autonomous workers in the construction sector. This recent development, stemming directly from the international research project presented in this book, indicates the added value of action research and how it can bring practical benefits.

Notes

Chapter opening image: A young worker in a brick factory in Colombia. Photograph courtesy of J. Maillard, ILO.

(1) The ABC forms part of the metropolitan region of São Paulo and groups seven municipalities: Santo André, São Bernardo do Campo, São Caetano do Sul, Diadema, Mauá, Ribeirão Pires and Rio Grande da Serra.

(2) Sistema de Contas Nacionais.

(3) Relacion Anual de Informaciones Sociales.

(4) Cadastro General de Empleados y Desempleados.

(5) Consolidação das Leis do Trabalho (CLT).

(6) Fundo de Garantia por Tempo de Serviço (FGTS).

(7) Norma Regulamentadora (NR).

(8) The FAT is financed by a 0.65% tax on revenues of private firms, 1% tax on revenues of public firms and 1% of costs of non-profit firms. It then pays UI the thirteenth wage (*abono salarial*), a fiscal stabilization fund and training initiatives from SENAI/SENAC and the National Development Bank (BNDES), which receives 40% of the FAT.

(9) More Equal Santo André.

(10) According to Denaldi, people living in a poverty situation are those living with a monthly income per capita of half the minimum salary (in 2000, this was RS240 or US$80). People living in a misery situation are those living under this poverty line.

(11) Premio Gestión Pública y Ciudadania.

(12) For the metropolitan region of São Paulo.

(13) At Santo André level.

7 Conclusions and Recommendations

Roderick Lawrence, Yves Flückiger, Cedric Lambert, Mariana Paredes Gil and Edmundo Werna

This concluding chapter presents some findings based on the analysis of the preceding chapters. The concept of decent work has been on the international labour agenda for nearly a decade. It has been debated in academic and professional forums. However, there is not much evidence to confirm the initial hypothesis that local authorities are playing a key role in implementing it in their municipalities via initiatives in the construction sector. Therefore, there still remains significant potential for innovation, despite the efforts that have been made and some good practices noted in this book. The guidelines and recommendations presented in this chapter are intended to help improve the dissemination and implementation of decent work at the national, and particularly at the local, levels in the immediate future.

7.1   General findings

Earlier chapters in this book clearly show that decent work remains a marginal concept, particularly at the local level. Recently, the ILO has made a major effort to implement the concept at the national level, and decent work country programmes have been designed throughout the world. However, application of the concept at the local level remains limited to a handful of cities and, to date, the ILO does not have a comprehensive plan to scale it up. The cases analyzed in this book have also shown that stakeholders at the local level generally lack a comprehensive understanding of the concept.

Since it was introduced by the ILO in 1999, decent work has been debated in academic and professional publications. This book confirms that there are numerous interpretations of it. In addition, several sets of indicators have been proposed to quantify decent work. However, these innovative contributions are not widely known. Consequently, a significant allocation of human and other resources is necessary for the concept of decent work to be better understood. In parallel, it is important to promote the concept for application at the local level.

The need for improved dissemination at the local level is illustrated by the case studies presented in this book. The concept of decent work was only known and applied in one of the three local authorities: Santo André. The field research found that both staff of the municipality of Santo André and representatives of workers unions were familiar with the concept of decent work, but none of the representatives of employers' associations or the working population knew about the concept. In the other selected cities, the concept of decent work was unknown.

There are indications that decent work may not have been applied at national and local levels for a number of reasons. First, decent work is a complex, multidimensional concept that includes four components and numerous sub-components. It is also a concept that is meant to deal with conventional sector based initiatives about employment generation, social

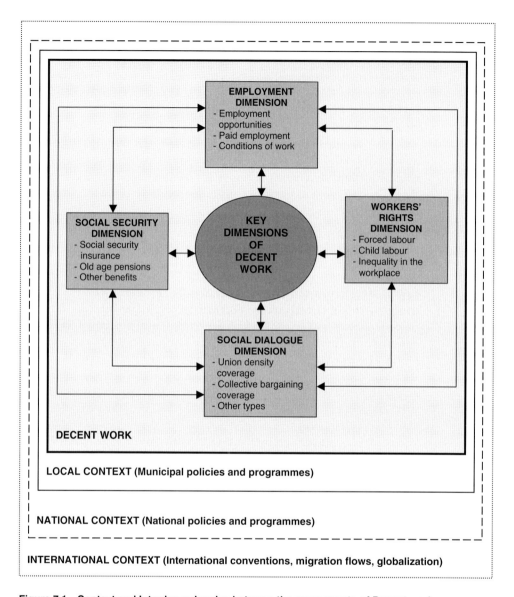

Figure 7.1   Context and Interdependencies between the components of Decent work

security, rights at the workplace and social dialogue. Hence it is meant to consider a vast number of conditions and requirements as well as the interrelations between them. The key question is whether this is feasible. This research project indicates that a positive answer to this question will depend on a number of contingent factors.

One contingent factor concerns current compartmentalized approaches that analyze working conditions. It is important to emphasize the idea that decent work is a multidimensional concept, as shown earlier in Figure 1.1.

In principle, unless the four key dimensions of decent work are considered simultaneously, the concept is not applied in its broadest sense. Therefore, and in order that decent work is effectively implemented, cross-sector co-operation is necessary. This required collaboration will imply challenges for actors and institutions at both the national and the local levels who commonly work in specific sectors. They will have difficulty dealing with the complexity of decent work until administrative reforms are instigated.

Another contingent factor is the principle that employment generation, social security, rights at the workplace and social dialogue need to be considered in terms of international conventions, national legislation and cultural customs. Hence, decent work is embedded in a broad institutional, political, economic and cultural context that ought to be understood at the local level by those actors and institutions who want to apply the concept. These contextual conditions have been explicitly considered and they are well illustrated by the case studies. Indeed, these contextual conditions are so important that the research team has decided to modify the conceptual framework proposed by Ghai (2005) that was reproduced in Figure 1.1. The new representation of decent work in Figure 7.1 clearly shows the four key components of decent work embedded in the local, national and international context.

Another important finding is that decent work should not be interpreted as a monolithic concept but as a multidimensional one that varies according to geographical and temporal factors. The four key components of decent work need to be considered in terms of the relative weightings of their subcomponents in precise localities. The case studies show how differences exist between countries and cities. This is an important finding that indicates that the subcomponents of decent work need to be assessed and understood by in-depth analytical studies at both the national and local levels. To date there are few contributions that address the need for methodological innovation. However, the implementation of decent work will depend on an accurate understanding of locality specific conditions which are embedded in a national and an international context. Therefore it is not unfair to claim that the research methodology that has been formulated and tested in this project can serve as a framework for future contributions.

This international research project has confirmed the pertinence of analyzing policies, programmes and projects in precise sectors at both national and local levels. This is an important conclusion of this project because the vast majority of contributions consider decent work only in terms of a broad economic and political framework without reference to sector based initiatives. The exceptions to this customary approach are a limited number of contributions that focus on agriculture. It is important to note that sector based contributions can be useful in the future to compare progress towards decent work in different sectors, and also highlight similarities and differences.

The set of indicators proposed to measure decent work was subdivided into the four components of decent work. The case studies validated this approach for the measurement of trends towards or away from decent work. The application of the set of decent work indicators in the case studies has shown that the selection of indicators was pertinent, and that these indicators were easy to identify in the different contexts at national and local levels for both the whole economy and the construction sector. In addition, they can be sourced from reliable quantitative databases (such as national surveys and censuses) and qualitative sources (questionnaires and interviews, for example).

The hypothesis of finding these indicators at the national and local levels for the whole economy and for the construction sector has been verified by the case studies. This approach permitted country and city level analyses and also enabled researchers to identify the role of the construction sector in the national economy. The selection of two time periods permitted analysis of the evolution of the socio-economic situation in the cities and countries. It is important to stress that this set of decent work indicators can be reapplied to other local authorities and countries.

Local governments, as well as other stakeholders, including workers' associations, cooperatives and community groups should explicitly be involved in policy implementation. Often the roles and responsibilities of these actors are underestimated. For example, the role of local authorities in the creation of employment, the procurement process and financing arrangements should not be forgotten when dealing with decent work. Although local authorities have rarely applied the concept of decent work, six examples of best practices are included at the end of Chapter 3. It is hoped local authorities will adopt a more proactive role in the near future.

7.2   Recommendations and guidelines

The following set of recommendations and guidelines are presented as one of the achievements of the international research project presented in this book.

Recommendation 1: actors and institutions involved in the promotion of decent work
Until now, the majority of contributions on decent work have been addressed to governments and the private sector at national level. These two audiences are important but limited. More concern should be given to identifying and working with all relevant actors involved in the promotion of decent work, bearing mind the importance of the local level. Local governments, as well as other stakeholders, including local worker' associations, cooperatives and community groups, are often overlooked but should not be neglected.

Public sector clients – for instance a ministry, a department of the central government, a local authority (regional, district or municipal) or even a community group – can create an adequate environment for successful project implementation.

Community groups – through community-based organizations (CBOs) – are becoming more and more important actors and are playing an increasing role in project identification, management, operation and maintenance with positive effect.

Contractors – central actors in the construction sector and in public procurement actions – can influence the attainment of social objectives included in the promotion of decent work. As responsible for the terms and conditions under which labour is engaged, they can enhance local involvement by subcontracting, and/or buying from local suppliers. They can also generate employment by using labour-based techniques.

We recommend the inclusion of all the different stakeholder groups involved in the generation of employment and the promotion of decent work, as shown by the case studies in Bulawayo and Santo André.

Guideline 1 All relevant actors involved in the promotion of decent work – national and local governments, community groups, cooperatives, workers, contractors and others – should be invited to participate in the decision-making and implementation processes. Both national and local platforms for dialogue between these actors and institutions can be proposed to facilitate communication and coordination.

Recommendation 2: emphasis on globalization and migration flows

The importance of taking into consideration not only the local but also the regional, national and international contexts should not be underestimated. Despite some recent contributions, there is still is a need for much more research on globalization, migration, urbanization and decentralization. These are relevant processes that ought to be understood when studying labour conditions in any locality.

Globalization is a phenomenon that influences countries all around the world. It has social, economic and political repercussions at local and national levels. The social and labour aspects of globalization are related to key dimensions of decent work, as shown by the case studies, particularly Bulawayo.

Globalization trends have been characterized by a greater integration of markets. Nevertheless, their impact on labour and population movements have been limited and regulated by severe immigration laws and policies. Globalization has important implications for international labour migration, acting as 'push and pull' factors. It has led to connections between labour markets. In parallel, globalization has also provoked great inconsistencies in employment opportunities, income and living standards.

In terms of migrant workers in the construction sector, several countries have acknowledged their dependence on such workers and tried to control and regularize the process of migration for the construction industry. However, there are still many people working illegally without any type of rights. Additionally, the recruitment process is often highly abusive and does not reflect the principles of decent work.

Guideline 2 Local conditions related to decent work should be analyzed bearing in mind regional, national and international processes. The context – not only local, regional and national but also international – should always be taken into consideration. Special attention should be given to cases where migration is significant in order to facilitate the integration and social welfare of migrant workers.

Recommendation 3: measurement of progress towards decent work
The measurement of progress towards decent work by the continual collection of quantitative and qualitative data is necessary in order to identify achievements and obstacles. The main problem hindering this objective is missing information at the local level. Commitment from national and local authorities is necessary to instigate and maintain a coordinated information system, which includes the set of indicators defined and validated during the research project presented in this book.

The categorization of the employed population is often based on traditional types of workers and does not take into consideration casual, temporary workers or informal workers. Reforms in the collection and analysis of data are necessary in order to monitor existing categories of the working population.

The aim of measuring decent work is to analyze labour using an individual approach rather than considering household income. We recommend median income as a more precise indicator than the poverty line, which is a more relative measurement.

It is important to distinguish clearly between the *relative* definition of the poverty line (e.g. 50% or 66.6% of the median income) and the *absolute* definition (e.g. US$2 per day). The absolute definition refers to levels of minimum income in order to satisfy basic needs (which is difficult to measure satisfactorily), while the relative definition refers to a lack in relation to the rest of the population (thus, even in a rich country, where every person's basic needs are met, a part of the population could still have a subjective sense of poverty in relation to others). This second concept is closely related to the notion of inequality.

Guideline 3 Collaboration between ILO offices, national statistics offices and national social security authorities should be strengthened, especially in regards to protocols for data collection and analysis in different localities. There is an urgent need to overcome missing information. In order to optimize the use of labour standards and decent work indicators, cities and local authorities should be consulted by ILO staff and, where necessary, training programmes should be provided.

Recommendation 4: from general discourse to practical implementation
Today there are too few incentives for local authorities to define and implement policies and projects that promote decent work. The ILO ought to be more proactive in promoting initiatives that could serve as beacons for change. Consequently, a large allocation of human and other resources is necessary in order to improve diffusion of the concept of decent work so that it can be much more widely applied.

In order for the decent work concept to become more widely known and more pertinent at the local level, we recommend that the ILO broaden its work beyond traditional stakeholders and expand its collaboration to more micro-level partners. The results of the Bulawayo case study suggest that the ILO should take a multi-level approach at the local level, working with National Employment Councils, local authority strategic teams, employers associations and employees unions.

Guideline 4 In order for the decent work concept to be better understood and more pertinent at the local level, it is imperative that the ILO adopt a strategic implementation plan, including an information and dissemination policy that broadens its work beyond traditional stakeholders and expands its collaboration to more micro-level partners.

Recommendation 5: benefits and challenges of decent work
The costs and investments necessary for the implementation of decent work in practice have not been considered. The benefits of decent work are rarely debated or publicized in regard to the local level, either by the ILO or/and national and local authorities. These benefits could be analyzed for target population groups at both national and local levels and from both short and long-term perspectives. There are several indications that the monetary and non-monetary investments necessary for the implementation of decent work need to be discussed further, including at the national level.

It is essential to distinguish economic and employment conditions in developed countries from those in transition and developing countries. In the case of developed countries, the costs and investments for the implementation of decent work will be smaller, as these countries already have a considerable legal structure and labour standards in place. For these reasons benefits of the implementation of decent work can be tabled. On the contrary, in transition and developing countries, the implementation of decent work will face more challenges. Costs and investment requirements will be higher because legal structures are often weak or inexistent and labour standards are not applied or do not reflect the reality of the majority of the working population who work as casual, informal or home workers. Consequently, in transition and developing countries, the benefits will only be recognized over a medium to long-term period. This is often the reason why public and private actors in transition and developing countries meet with difficulties in implementing decent work; it takes too much time, it is expensive and it does not immediately resolve the problems of the manual worker population.

Guideline 5 There is an urgent need to monitor the human and financial resources required to implement decent work in specific localities, and the positive outcomes of these investments. The ILO should provide a protocol for monitoring projects and policies at the local level in countries with different economies in order to achieve this objective in a systematic way.

Recommendation 6: clarification of conventional categories: the case of temporary workers

Research and practice on labour issues tend to rely heavily on the categories of formal and informal workers. Whilst the informal sector remains important (see next recommendation), overemphasis on this formal-informal dichotomy does not capture structural changes in the world of labour, such as the occurrence of large numbers of temporary workers. This has consequences for decent work and needs to be addressed.

First, it is important to explore the possibilities of social security coverage for temporary workers through existing social security schemes. There is evidence from many countries that employers do not contribute to social security funds for those workers who are on temporary contracts. Hence the workers who are most in need receive no health care, no holiday pay and no protection against loss of pay when they are unable to work due to unemployment, ill-health, accidents or old age. Many of the self-employed (without employees) are believed to be actually working for employers, but without social security contributions. Similarly, most workers on temporary contracts are not entitled to benefits during periods of unemployment between contracts.

When full social protection is not feasible because of limited financial capacity, we recommend that priority be given to needs which are most pressing for the workers concerned. It is essential to have specific health and safety proposals at the design stage in order to identify possible risks and how to deal with them. In the case of serious accidents, when it is not possible to include casual unskilled workers under the official social security system, accident insurance will need to be provided to pay for disability, death, serious medical expenses and loss of income. The best option for dealing specifically with minor accidents in community contracting may be the creation of some form of local fund that can respond quickly to meet medical expenses.

Second, regarding social dialogue, it is important to recognize local informal groups that represent workers (such as NGOs, CBOs and workers groups who can develop home-grown solutions). Another important point is to discuss mechanisms for implementing labour standards in advance of work. Social dialogue with employers, and also with governments, has traditionally been a powerful means for workers to bargain collectively for better wages and improved working conditions. However, the vast numbers of temporary, casual, informal and unemployed workers today find it very difficult to organize themselves to engage in social dialogue.

The case study in Dar es Salaam explicitly emphasized the high number of casual workers, namely in the construction sector, and the importance of considering them in the implementation of policies and projects to promote decent work.

Guideline 6 Specific policies should be developed and strategic actions should be applied in order to deal with temporary and casual workers in the promotion of decent work. Reforms in data collection are necessary.

Recommendation 7: the importance of the informal sector

During recent decades, the informal sector has been growing all around the world and the majority of new employment in the last decade has been in the informal economy. This situation is accentuated in developing and transition countries. Therefore, the increase of the informal sector has been related not only to the capacity of formal firms to absorb labour but also to their willingness to do so. More firms are decentralizing production and organizing work through flexible specialized production units, some of which remain unregistered or informal. As part of their efforts to improve competitiveness, firms are increasingly operating with a small core of wage employees with regular terms and conditions based in a fixed formal workplace and a growing periphery of atypical and often informal workers in different types of workplaces dispersed in different locations. These measures often include outsourcing or subcontracting and a shift away from regular employment relationships to more flexible and informal employment relationships.

In the developing world, where the informal sector is very important, protection against social risks has been provided by the family or community. Where the formal and traditional systems are inexistent, several examples of informal social insurance mechanisms based on principles of solidarity and reciprocity occur. In recent years, various groups of informal workers have set up their own micro-insurance schemes. One option is to form organizations among workers, which enables them to achieve various objectives, including stronger negotiating power with regard to the government as well as public and private health providers, sharing of knowledge and greater financial stabilization through reinsurance. A second option is to allocate more effort to the promotion of micro-insurance, since a large part of the target population may still not be aware of the benefits of such a scheme.

We recommend that local authorities play an important role in setting up area-based social protection schemes and working in partnership with local groups, CBOs and civil society.

Guideline 7 Workers in the informal sector should be taken into consideration as a category of all employed persons. Social security, safety and health insurance systems should be implemented for this group of workers, as has been done for those employed in the formal sector.

Recommendation 8: lack of application of the labour legislation

Instability of employment is one of the major problems in the construction sector. Fluctuations in demand, the project base of construction and the widespread use of contracting all conspire to make it difficult for contractors to obtain a steady flow of work which would allow them to provide continuity of employment. Hence, there is a constant friction between employers' needs for 'flexibility' and workers' needs for stable jobs. Construction workers are conventionally employed on a short-term basis, for the duration of the whole or part of a project, which means no guarantee of future work. This results in an increasing number of temporary and informal workers. Therefore, we recommend that labour legislation is also applied to temporary and informal workers.

The problem in the construction sector is not a lack of labour law, which exists at the national and local levels. The real problem is the lack of efficient instruments to ensure that labour laws are applied and monitored. Contracts in the construction sector can be used and important instruments for the implementation of labour standards can be enforced. It is useful to show the benefits of these standards not only for workers, but also for employers and contractors.

Guideline 8 Legal and administrative measures should be implemented to ensure that all rights and responsibilities in labour laws are applied. Monitoring and surveillance are essential and sanctions should be enforced when legislation is not respected.

Recommendation 9: procurement initiatives

Procurement can play an important role in the implementation of decent work and the achievement of social objectives. Targeted procurement is a system for awarding tenders that provides the option to set goals for achieving contractually enforceable socio-economic objectives, retaining donor rules of competition, fairness, efficiency and transparency. Procurement initiatives have been used to support local economic development, promote growth within the small business sector and target the unemployed through poverty alleviation. Nevertheless, it is important to have appropriate planning and design, clearly identify goals and possess the ability and willingness to apply sanctions.

Frequently, labour-standards in contracts are not implemented because they are not monitored or enforced through incentives and/or sanctions. Two Swiss examples, presented in Boxes 7.1 and 7.2, illustrate incentives aimed to enforce the implementation of labour standards; the first uses procurement whereas the second is related to certification. The Certification of Equal Compensation for Women and Men is a Swiss initiative to enforce equality of salary between men and women in the private sector. The Federal Law for Public Markets prescribes a legal framework for Swiss public markets which respected labour standards, namely workers' rights.

7.3   Conclusion

During the last decade, policies and projects that counteract increasing levels of unemployment and human vulnerability (especially the precarious nature of an increasing share of paid work) have become a high priority for some international organizations –including the International Labour Organization – many national and regional governments in both developed and developing countries, and local authorities. Therefore, the principles of decent work are an important target for institutions and actors at the international, national and local levels.

In the context of a globalizing economy, it has been increasingly recognized that local authorities provide the main location, infrastructure, community services and workforce for a growing share of the emerging global economy as well as other economic sectors. Cities and municipalities are key actors in the generation of Gross Domestic Product. Therefore, local authorities can be the catalyst for the promotion of decent work that guarantees income, equity, security and dignity to members of local communities. They can also be the arena for decent work, living conditions, social security and dialogue between members of civil society. This role has rarely been adopted and the example of Santo André is a notable exception. It is hoped that other local authorities will follow the examples of best practice included in earlier chapters.

This book has clearly shown that much more needs to be done at the global, national and particularly local levels. The principles of decent work should not be considered as constraints but as the catalyst for change towards better living and working conditions. It is hoped that this book has presented avenues for the improved dissemination and implementation of decent work in order to achieve this objective.

Box 7.1   Federal Law on Public Procurement in Switzerland

The Federal Assembly of the Swiss Confederation adopted the Federal Law on Public Procurement in December 1994 and this entered into force on 1 January 1996.

The Federal Law on Public Procurement ensures that Swiss federal authorities respect certain principles in regards to potential suppliers in all tendering procedures.

According to the law (Article 8), when initiating public procurement, a Swiss authority is required to insure that the principles related to worker protection and working conditions are respected by potential suppliers (Al. 1, lit. b). The Swiss authority must also guarantee that the potential suppliers respect equal salary treatment between women and men (Al. 1, lit. c).

The Federal law also specified that the Swiss authority can control, or make a third party control, the implementation and respect of the obligations cited above by the supplier. Upon request, the supplier shall prove the fulfilment of its obligations (Al. 2).

Non-fulfilment of the above-mentioned obligations by a supplier allows the Swiss authority to exclude this supplier from tendering procedures or to revoke the contract if it has been awarded (Art. 11, lit. d).

'Art. 8: Principes et conditions de participation:

(1) Les principes ci-après doivent être observées lors de la passation de marchés publics: . . . (b) pour les prestations fournies en Suisse, il n'adjuge le marché qu'à un soumissionnaire observant les dispositions relatives à la protection des travailleurs et les conditions de travail . . . (c) il n'adjuge le marché qu' à un soumissionnaire garantissant à ses salariés l'égalité de traitement entre femmes et hommes, sur le plan salarial, pour les prestations fournies en Suisse.

(2) L'adjudicateur est en droit de contrôler ou de faire contrôler l'observation des dispositions relatives à la protection des travailleurs, aux conditions de travail et de l'égalité de traitement entre hommes et femmes. Sur demande, le soumissionnaire doit apporter la preuve qu'il les a respectées.

Art 11: Exclusion de la procédure et révocation de l'adjudication:

L'adjudicateur peut révoquer l'adjudication ou exclure certains soumissionnaires de la procédure ainsi que les rayer de la liste . . . notamment lorsque:

(d) Ils ne satisfont pas aux obligations fixées à l'article 8.

Source: Swiss Confederation, Federal Law on Public Procurement http://www.admin.ch/ch/f/rs/c172_056_1.html (consulted on 5 April 2007).

> **Box 7.2   Certification of Equal Compensation for Women and Men, Switzerland**
>
> Many private sector enterprises in Switzerland claim to have a salary policy which respects equality between men and women. On the other hand, figures published by the Swiss Federal Statistical Office show a difference of more than 20% between the salaries earned by women and men. The equal-salary certification allows companies to test their equal compensation policies in total confidentiality.
>
> This initiative is a partnership between Véronique Goy Veenhuys, founder and coordinator of equal-salary, and Yves Flückiger, director of the Employment Observatory (OUE) at the University of Geneva. Professor Flückiger has developed an equation of salaries, a scientific tool which has been approved by the Swiss Federal Court in its decision of 2003. Following this decision, several major Swiss corporations have asked the OUE to evaluate their compensation policies, expressing both a concern and a need.
>
> By becoming equal-salary certified, an enterprise has proof that it respects equal compensation for men and women and can communicate openly about its salary policy. By defusing a source of conflict, the enterprise can directly influence employee motivation and its ability to attract the best female candidates. Equal-salary helps to improve the corporation's image both internally and externally with customers, suppliers and other stakeholders.
>
> By making it part of their corporate strategy, the equal-salary certification helps enterprises to improve their financial performance.
>
> The Swiss Confederation financially supports the project through funds allocated by Swiss law to the equality of men and women.
>
> The Equal-Salary Certification Board has also been established and is comprised of prominent Swiss women and men who support equality in general and the equal-salary in particular. As noted by the Advisory Board: 'Equal salaries for men and women for equal work' must become a reality in Switzerland. Thanks to its close collaboration with enterprises committed to putting this concept into practice, equal-salary is certainly an initiative that contributes to this aim.
>
> *Source*: http://www.equal-salary.ch (consulted on 20 August 2008).

Chapter opening image: A woman carrying a basket of stones weighing more than 20 kg on her head, on a construction site in Hyderabad, India. Photograph courtesy of Marcel Crozet, ILO.

References

Abramo, L. (ed.) (2006) *Trabajo Decente y Equidad de Género en América Latina.* Oficina Internacional del Trabajo, Santiago.

Anker, R., Chernyshev, I., Egger, P., Mehran, F. & Ritter, J. (2002) *Measuring Decent Work with Statistical Indicators.* Working Paper No. 2. Policy Integration Department, International Labour Organization, Geneva.

Anker, R., Chernyshev, I., Egger, P., Mehran, F. & Ritter, J. (2003) Measuring decent work with statistical indicators. *International Labour Review,* **142** (2) (Special Issue: Measuring Decent Work), 147–178.

Anon (2002) *Annual Report of the State of Corruption in Tanzania.* Tanzania Development Gateway: www.tanzaniagateway.org

Baccaro, L. (2001) *Civil Society, NGOs, and Decent Work Policies: Sorting out the Issues.* International Institute for Labour Studies, International Labour Organization, Geneva.

Bakker, S., Kirango, J. & van der Rees, K. (2000) Both sides of the bridge: public-private partnership for sustainable employment in waste management, Dar es Salaam, paper submitted for ILO workshop, Manila, 18–21 September. In J. Plummer (ed.) (2002) *Focusing Partnerships: a Sourcebook for Municipal Capacity Building in Public-private Partnerships.* Earthscan Publications, London.

Barros Silva, P.L., Pochmann, M. & Abrahão, J. (1999) *Ação dos Atores na Superação da Exlusõ Social no Brasil: Experiéncias Autais.* Equipo Técnico Multidisciplinario para Argentina, Brasil, Chile, Paraguay y Uruguay, International Labour Organization, Geneva.

Barten, F., Mitlin, D., Mulholland, C., Hardoy, A. & Stern, R. (2006) *Healthy Governance/Participatory Governance. Towards Integrated Approaches to Address the Social Determinants of Health for Reducing Health Inequity.* Background paper for the knowledge network on urban settings of the WHO CDSH, WHO Kobe Centre, Japan.

Barten, F., Sousa Santana, V., Rongo, L., Varillas, V. & Pakasi, T.A. (2008) Contextualising workers' health and safety in urban settings: the need for a global perspective and an integrated approach. *Habitat International,* **32** (2), 223–236.

Batley, R. (1998) *Urban Water in Zimbabwe: Performance and Capacity Analysis.* International Development department, University of Birmingham.

Bazan, L. & Schmitz, H. (1997) *Social Capital and Export Growth: an Industrial Community in Southern Brazil.* Discussion Paper 361. Institute of Development Studies, Brighton.

Bescond, D., Chataignier, A. & Mehran, F. (2003) Seven indicators to measure decent work: an international comparison. *International Labour Review,* **142** (2) (Special Issue: Measuring Decent Work), 179– 212.

Bond, P. & Manyanya, M. (2002) *Zimbabwe's Plunge: Exhausted Nationalism, Neo-liberalism and the Struggle for Social Justice.* Merlin, London; Weaver Press, Harare.

Bonnet, F., Figueroa, J.B. & Standing, G. (2003) A family of decent work indexes, *International Labour Review*, **142** (2) (Special Issue: Measuring Decent Work), 213– 238.

Bourdillon, M. (2006) Children and work: a review of current literature and debates. *Development and Change*, **37** (6), 1201–1226.

Boyer, R. (2006) *Employment and Decent Work in the Era of 'Flexicurity'.* UN/DESA, DESA Working Paper No. 32.

Bulawayo City (1994) *Report of Interdepartmental Committee on Incentives for Development.* Projects Committee Team, City of Bulawayo.

Bulawayo City (1995) *Criteria for Nomination and Selection of a Winning Local Authority.* Preparation for Habitat II conference, City of Bulawayo.

Bulawayo City (1999) *Outstanding Development of Virgin Land Offered to Private Contractors for Residential Development.* Town Lands and Planning Report, City of Bulawayo.

Bulawayo City (2000) *Bulawayo Master Plan 2000–2015: the Written Statement.* City Council, City of Bulawayo.

Cacciamali, M.C., de Lima Becerra, L., do Vale Souza, A., Mello, R. & Saboia, J. (1998) *Desafios da Modernização e Setor Infromal Urbano: o Caso de Brasil.* Oficina Regional de la OIT para America Latina y el Caribe, Organización Internacional del Trabajo, Lima.

Carroll, R. (2006) As Zimbabwe's economy collapses, a tiny few make huge profits. *The Guardian*, 25 April, London, p. 21.

CB Richard Ellis (2006) *Report on the Harare Property Market: the First Quarter of 2006.* CBRE, Harare.

Chandler, J. & Lawless, P. (1985) *Local Authorities and the Creation of Employment.* Gower Publishing Company Limited, Aldershot.

Chernyshev, I. (2003) Decent work statistical indicators: strikes and lockouts statistics in the international context. *Bulletin of Labour Statistics*, Policy Integration Department, International Labour Organization, Geneva, pp. ix–xxxviii.

Chipika, J.T. (2005) *Zimbabwe Country Situation Analysis Report to Inform the Formulation of a Decent Work Country Programme (DWCP) for Zimbabwe.* Final draft report prepared for International Labour Organization (ILO-SRO), Harare.

Christensen, P. & O'Brien, M. (eds) (2003) *Children in the City: Home, Neighborhood and Community.* Routledge, London.

Clifton, J. & van Esch, W. (2000) *Hanna Nassif Urban Upgrading Project Phase II, Dar es Salaam, Tanzania: Study on Implementation of Community Managed and Labour Based Works Using Community and Private Contracts: Consultancy Report.* International Labour Organization, Geneva.

Confederation of International Contractor's Associations (CICA) (2002) *Industry as a Partner for Sustainable Development: Construction.* Confederation of International Contractor's Associations, Paris.

Contractors Registration Board (CRB) (2001) *Report on Baseline Study on Safety and Health Conditions in Construction Sites in Tanzania.* CRB, Dar es Salaam.

Coulson, A. (1982) *Tanzania: a Political Economy.* Clarendon Press, Oxford.

CSO (1998) *Zimbabwe Quarterly Digest of Statistics.* Central Statistical Office (CSO), Government of Zimbabwe, Harare (4 December, p. 11).

CSO (2000) *1999 Indicator Monitoring Labour Force Survey.* Central Statistical Office, Harare.

CSO (2001) *Zimbabwe Compendium of Statistics, 2000.* Central Statistical Office, Harare.

CSO (2002) *Census 2002 National Report.* Central Statistical Office, Harare.

CSO (2003) *Zimbabwe Quarterly Digest of Statistics.* Central Statistical Office, Government of Zimbabwe, Harare (1–4 March to December, p. 11).

CSO (2004) *Zimbabwe Labour Statistics.* Central Statistical Office, Harare.

Denaldi, R. (1997) Viable self-management: the FUNACOM housing program of the São Paulo municipality. *Habitat International,* **21** (2), 213–227.

Denaldi, R. (2003) *Politicas de urbanização de favelas: evolução e impasse.* PhD thesis. University of São Paulo, Faculty of Architecture and Urban Planning, FAUUSP, São Paulo.

Denaldi, R. (2004) *Santo André: Integración de Programas para Promover la Inclusión Social.* Ayuntamiento de Santo André, Santo André.

Dev, S.M. (1995) India's (Maharashtra) Employment Guarantee Scheme: lessons from long experience. In J. von Braun (ed.) *Employment for Poverty Reduction and Food Security.* IFPRI, Washington, pp. 108–143.

Diário do Grande ABC (2006) Trabalho infantil cresce 10.3% no país. *Diário do Grande ABC.* Sábado 16 de setembro, p. 10, São Paulo.

DIESSE (2001) *Os Trabalhadores e a Restructuração Produtiva na Construção Civil Brasileira.* Estudos Setoriales DIESSE No. 12. Departamento Intersindical de Estatística e Estudos Sócio-econômicos, DIESSE, São Paulo.

DIESSE (2004) *Caracterização do Setor da Construção Civil.* Estudos e Pesquisas, Departamento Intersindical de Estatística e Estudos Sócio-econômicos, DIESSE, São Paulo.

Durand, J.P. (1998) Is the 'better job' still possible? *Today's Economic and Industrial Democracy,* **19** (1), 185–198.

Dworkin, R. (2006) It is absurd to calculate human rights according to a cost benefit analysis. Comment and Debate. *The Guardian,* 24 May, London, p. 28.

Economist Intelligence Unit (EIU) (2003) *Zimbabwe Country Profile.* The Economist Intelligence Unit, London.

Economist Intelligence Unit (EIU) (2005) *Zimbabwe Country Profile.* The Economist Intelligence Unit, London.

ECOSOC (2006a) *Globalization and Global Migration.* United Nations Economic and Social Council, Geneva.

ECOSOC (2006b) *Full and Productive Employment and Decent Work: Dialogues at the Economic and Social Council.* Department of Economic and Social Affairs – Office for ECOSOC Support and Coordination, United Nations, New York.

Edmonds, G.A. (1979) Construction industry in developing countries. *International Labour Review,* **118** (3), 335–369.

Education International (2004) *Barómetro de la IE sobre los Derechos Humanos y Sindicales en el Sector de la Educación 2004: la Educación es un Derecho Humano.* Education International, Brussels.

Egger, P. (2002) Towards a policy framework for decent work. *International Labour Review,* **141** (1–2), 161–174.

Egger, P. & Sengenberger, W. (2001) *Decent Work: Issues and Policies.* Eldis & ILO, Montevideo: http://www.eldis.org

Eicher, C.K. (2003) *Flashback: Fifty Years of Donor Aid to African Agriculture.* Conference Paper No. 16: paper presented at the IFRPI, NEPAD, CTA conference on 'Successes in African Agriculture', Pretoria.

Fields, G.S. (2003) Decent work and development policies. *International Labour Review,* **142** (2) (Special Issue: Measuring Decent Work), 239–262.

Fjeldstad, Odd-Helge (2004) *Trust in Public Finance: Citizens Views on Taxation by Local Authorities in Tanzania.* Formative process research on local government reform in Tanzania, Project Brief LGR 12: www.tanzaniagateway.org

Frota, L. (2008) Securing decent work and living conditions in low-income urban settlements by linking social protection and local development: a review of case studies. *Habitat International,* **32** (2), 203–222.

Gabre-Madhin, E.Z. & Haggblade, S. (2004) Successes in African agriculture: results of an expert survey. *World Development,* **32** (5), 745–766.

Gaude, J. & Watzlawick, H. (1992) Employment creation and poverty alleviation through labour-intensive public works in least developed countries. *International Labour Review,* **131** (1), 3–18.

Ghai, D. (2002) *Decent Work: Concepts, Models and Indicators.* Discussion Paper No. 139. International Institute of Labour Studies, International Labour Organization, Geneva.

Ghai, D. (2003) Decent work: concept and indicators. *International Labour Review,* **142** (2) (Special Issue: Measuring Decent Work), 113–146.

Ghai, D. (2005) *Decent Work: Universality and Diversity.* Discussion Paper Number 159. International Institute of Labour Studies, International Labour Organization, Geneva.

Ghai, D. (ed.) (2006) *Decent Work: Objectives and Strategies.* International Institute for Labour Studies, International Labour Organization, Geneva.

Godfrey, M. (2003) *Employment Dimensions of Decent Work: Trade-offs and Complementarities.* Discussion Paper No. 148. International Institute of Labour Studies, International Labour Organization, Geneva.

GoZ (1976) *Factories and Works Regulations.* Rhodesia Government Notice, No. 263 (Cap. 28). Government of Zimbabwe, Harare.

GoZ (1982) *The 1982–1985 Transitional National Development Plan.* Government of Zimbabwe, Harare.

GoZ (1990a) *National Social Security Authority (Accident Prevention and Workers' Compensation Scheme) Notice.* Statutory Instrument 68 of 1990. Government of Zimbabwe, Harare.

GoZ (1990b) *Collective Bargaining Agreement: Engineering and Iron and Steel Industry.* Statutory Instrument 282 of 1990 (Act 16/85). Government of Zimbabwe, Harare.

GoZ (1990c) *Labour Relations Regulations.* Statutory Instrument 31 of 1993 (Act 16/8). Government of Zimbabwe, Harare.

GoZ (1990d) *Collective Bargaining Agreement: Engineering and Iron and General Engineering Section.* Statutory Instrument 100 of 1997 (Cap. 28: 01). Government of Zimbabwe, Harare.

GoZ (1993) *Preference for Indigenous Building Contractors.* Treasury Circular No. 2 of 1993, Ministry of Finance, Causeway. Government of Zimbabwe, Harare.

GoZ (1996a) *National Social Security Authority Act* (Cap. 17: 04). Government of Zimbabwe, Harare.

GoZ (1996b) *The Regional Town and Country Planning Act* (revised edition). Government of Zimbabwe, Harare.

GoZ (1996c) *Labour Relations Act* (Cap. 28: 01) (revised edition). Government of Zimbabwe, Harare.

GoZ (1996d) *Factories and Works Act* (Cap. 14: 08) (revised edition). Government of Zimbabwe, Harare.

GoZ (1996e) *War Victims and Compensation Act* (Cap. 11: 16) (revised edition). Government of Zimbabwe, Harare.

GoZ (1999) *Collective Bargaining Agreement: Construction Industry.* Statutory Instrument 244 of 1999 (Cap. 28: 01). Government of Zimbabwe, Harare.

GoZ (2000) *National Housing Policy for Zimbabwe.* Ministry of Public Construction and National Housing, Government of Zimbabwe, Harare.

GoZ (2002a) *Labour Relations Amendment: to Amend Labour Relations Act* (Cap. 28: 01) *and the Public Service Act* (Cap. 16: 04). Government of Zimbabwe, Harare.

GoZ (2002b) *Public Order and Security Act* (Cap. 11: 17), No. 1. Government of Zimbabwe, Harare.

GoZ (2002c) *Access to Information and Protection of Privacy Act* (Cap. 10: 27), No. 5. Government of Zimbabwe, Harare.

GoZ (2005a) *Labour Amendment Act: Labour Act* (Cap. 28: 01). Government of Zimbabwe, Harare.

GoZ (2005b) *Constitution of Zimbabwe* (as amended in September 2005). Legal Resources Foundation, Government of Zimbabwe, Harare.

GoZ (2005c) *Response by the Government of Zimbabwe to the Report by the UN Special Envoy on Operation Murambatsvina.* Government of Zimbabwe, Harare.

GoZ (no date) *Promoting Women, Employment, Gender Equality and Equity.* Ministry of Women Affairs, Gender and Community Development, Government of Zimbabwe, Harare.

GoZ/ILO (2006) *Decent Work Country Programme.* Government of Zimbabwe and International Labour Organization, Harare.

GoZ/USAID (1994), *Zimbabwe Private Sector Housing Programme Monitoring and Evaluation System: Baseline Surveys and Findings.* Government of Zimbabwe and United States Agency for International Development, Harare.

GoZ/USAID (1996) *Gender Analysis of The Zimbabwe Housing Guaranty Program.* Prepared by Plan Inc. Zimbabwe Pvt (Ltd) for United States Agency for International Development Mission in Zimbabwe/Ministry of Public Construction and National Housing, Harare.

Haas, F., Oliveira Muniz, J. & de Oliveira Lima, J. (2003) *Brazil Child and Adolescent Domestic Work in Selected Years from 1992 to 1999: a National Report.* International Programme on the Elimination of Child Labour (IPEC). International Labour Organization, Geneva.

Halla, F. (1994) Coordinating and participatory approach to managing cities: the case of the sustainable Dar es Salaam project in Tanzania. *Habitat International*, **18** (3), 19–31.

Halla, F. & Majani, B. (1999) The environmental planning and management process and the conflict over outputs in Dar es Salaam. *Habitat International*, **23** (3), 339–350.

Hamilton, M. & Ndubiwa, M. (eds) (1994) *Bulawayo: a Century of Development – 1984 1994*. The Argosy Press, Harare.

Hawkins, J., Herd, C. & Wells, J. (2006) *Modifying Infrastructure Procurement to Enhance Social Development*. Engineers Against Poverty (EAP) and Institution of Civil Engineers (ICE), London.

Helmsing, B. (2001) *Partnerships, Meso-institutions and Learning. New Local and Regional Economic Development Initiatives in Latin America*. Institute of Social Studies, The Hague.

Hepple, B. (2003) *Rights at Work*. Discussion Paper No. 147. International Institute for Labour Studies, International Labour Organization, Geneva.

IBGE (1992) *Pesquisa Nacional por Amostra de Domicílios – PNAD 1992*. Instituto Brasileiro de Geografia e Estatística, Rio de Janeiro.

IBGE (2000) *Pesquisa Mensual de Emprego: Janeiro de 2000*. Indicadores IBGE, Instituto Brasileiro de Geografia e Estatística, Rio de Janeiro.

IBGE (2001a) *Pesquisa Nacional por Amostra de Domicílios – PNAD 2001*. Instituto Brasileiro de Geografia e Estatística, Rio de Janeiro.

IBGE (2001b) *Mapa do Mercado de Trabalho no Brasil: 1992–1997*. Instituto Brasileiro de Geografia e Estatística, Rio de Janeiro.

IBGE (2002) *Pesquisa Anual da Indústria da Construção 2002, Volume 12*. Instituto Brasileiro de Geografia e Estatística, Rio de Janeiro.

IBGE (2003) *Pesquisa Nacional por Amostra de Domicílios – PNAD 2003*. Instituto Brasileiro de Geografia e Estatística, Rio de Janeiro.

IBGE (2005) *Pesquisa Nacional por Amostra de Domicílios – PNAD 2005*. Instituto Brasileiro de Geografia e Estatística, Rio de Janeiro.

ILO (1996) *The Future of Urban Employment*. Report and Proceedings of the Urban Employment Dialogue for the Twenty-first Century, held during the Second United Nations Conference on Human Settlements (Habitat II), Istanbul.

ILO (1998) *The Future of Urban Employment*. International Labour Organization, Geneva.

ILO (1999a) *Decent Work*. International Labour Conference, Eighty-seventh Session, Report of the Director-General, International Labour Organization, Geneva.

ILO (1999b) *Employment-intensive Investment in Infrastructure: Jobs to Build Society*. International Labour Organization, Geneva.

ILO (2000a) *Decent Work for Women: an ILO Proposal to Accelerate the Implementation of the Beijing Platform for Action*. Bureau for Gender Equality, International Labour Organization, Geneva.

ILO (2000b) *Perspectives on Decent Work: Statements by the ILO Director-General*. International Labour Organization, Geneva.

ILO (2000c) *Seguridad y Salud en el Trabajo de Construcción: el Caso de Bolivia, Colombia, Ecuador y Perú*. Equipo Multidisciplinario para los Países Andinos,

Oficina Subregional para los Países Andinos, Organización Internacional del Trabajo, Lima.

ILO (2000d) *Employment-intensive Investment in Infrastructure: Jobs to Build Society*. International Labour Organization, Geneva.

ILO (2001a) *Panorama Laboral 2001 América Latina y el Caribe*. Oficina Regional para América Latina y el Caribe, Organización Internacional del Trabajo, Lima.

ILO (2001b) *Reducing the Decent Work Deficit: a Global Challenge*. International Labour Conference, Eighty-ninth Session, Report of the Director-General, International Labour Organization, Geneva.

ILO (2001c) *The Construction Industry in the Twenty-first Century: its Image, Employment Prospects and Skill Requirements*. Tripartite Meeting on the Construction Industry in the Twenty-first Century: its Image, Employment Prospects and Skill Requirements, International Labour Organization, Geneva.

ILO (2002a) *Decent Work and the Informal Economy*. Report IV. International Labour Conference Ninetieth Session, International Labour Organization, Geneva.

ILO (2002b) *Mondialisation et Travail Décent dans les Amériques*. Rapport du Directeur Général, Organisation International du Travail, Genève.

ILO (2002c) *Panorama Laboral 2002 América Latina y el Caribe*. Oficina Regional para América Latina y el Caribe, Organización Internacional del Trabajo, Lima.

ILO (2002d) *Every Child Counts: New Global Estimates on Child Labour*. International Programme on the Elimination of Child Labour (IPEC), Statistical Information and Monitoring Programme on Child Labour (SIMPOC), International Labour Organization, Geneva.

ILO (2003a) *Decent Work in Agriculture*. International Workers' Symposium on Decent Work in Agriculture, Bureau for Workers' Activities, International Labour Organization, Geneva.

ILO (2003b) *Working out of Poverty*. Report of the Director-General, International Labour Conference, Ninety-first Session, International Labour Organization, Geneva.

ILO (2003c) *Time for Equality at Work*. Global Report under the follow-up to the ILO Declaration on Fundamental Principles and Rights at Work, Report of the Director-General, International Labour Conference Ninety-first Session, International Labour Organization, Geneva.

ILO (2003d) *Brazil Child and Adolescent Domestic Work in Selected Years from 1992 to 1999: a National Report*. International Programme on the Elimination of Child Labour (IPEC), International Labour Organization, Geneva.

ILO (2004a) *Cities at Work: Employment Promotion to Fight Urban Poverty*. International Labour Organization, Geneva.

ILO (2004b) *Achieving Decent Work by Giving Employment a Human Face*. IFP/Dialogue, Paper No. 7, International Labour Organization, Geneva.

ILO (2005a) *Panorama Laboral 2005 América Latina y el Caribe*. Oficina Regional para América Latina y el Caribe, Organización Internacional del Trabajo, Lima.

ILO (2005b) *Baseline Study of Labour Practices on Large Construction Sites in Tanzania*. Working Paper 225, Sectoral Activities Department, International Labour Organization, Geneva.

ILO (2005c) *The Nexus of Growth, Investment and Jobs*. Report of the Tripartite Workshop for the Southern Africa Region (5–6 December), Johannesburg.

ILO (2005d) *ILO Decent Work Country Programmes: a Guide Book*. International Labour Organization, Geneva.

ILO (2005e) *Global Employment Trends 2005*. International Labour Organization, Geneva.

ILO (2006a) *Decent Work in a Global Economy*. International Institute for Labour Studies, International Labour Organization, Geneva.

ILO (2006b) *Trabajo Decente en las Américas: una Agenda Hemisférica, 2006–2015*. Informe del Director General, XVI Reunión Regional Americana, Organización Internacional del Trabajo, Brasilia – mayo 2006. International Labour Organization, Geneva.

ILO (2006c) *Agenda Nacional de Trabalho Decente*. Organización Internacional del Trabajo, Brasilia.

ILO (2006d) *Promoting Decent Work in the Americas: Hemispheric Agenda 2006–2015*. International Labour Organization, Geneva.

ILO (2006e) *A Strategy for Urban Employment and Decent Work*. International Labour Organization, Geneva.

ILO (2007a) *Toolkit for Mainstreaming Employment and Decent Work*. International Labour Organization, Geneva.

ILO (2007b) *Decent-work and Poverty Reduction Strategies*. International Labour Organization, Geneva.

ILO (2007c) *Decent Work for Sustainable Development*. International Labour Conference, Ninety-sixth Session, Report of the Director-General, International Labour Organization, Geneva.

ILO (2007d) *The Decent Work Agenda in Africa 2007–2015*. Report of the Director-General, Eleventh African Regional Meeting, International Labour Organization, Geneva.

ILO (2008a) *Integrating Local Economic Development and Social Protection: Experiences from South Africa*. STEP Working Paper, Fighting Social Exclusion Series, International Labour Organization, Geneva.

ILO (2008b) *Local Development, Social Protection and Inclusion: Typology of Selected Initiatives in Brazil*. STEP Working Paper, Fighting Social Exclusion Series, International Labour Organization, Geneva.

ILO (2008c) *Global Employment Trends*. International Labour Organization, Geneva.

Intermarket Research Economics (2004) *Confederation of Zimbabwe Industries (CZI): State of the Local Manufacturing Sector in 2003*. Intermarket Research/Report for CZI, Harare.

International Union of Local Authorities (1971) *Local Government as Promoter of Economic and Social Development*. International Union of Local Authorities, The Hague.

International Union of Local Authorities (1975) *Urbanization Today*. International Union of Local Authorities, The Hague.

Jason, A. (2005) *Informal Construction Workers in Dar es Salaam, Tanzania.* Working Paper 226, Sectoral Activities Department, International Labour Organization, Geneva.

Jason, A. (2008) Organizing informal workers in the urban economy: the case of the construction industry in Dar es Salaam. Tanzania. *Habitat International,* **32** (2), 192–202.

Jose, A.V. (ed.) (2002) *Organized Labour in the Twenty-first Century.* International Institute for Labour Studies, International Labour Organization, Geneva.

Kantor, P. (2008) Diversification and security? Labour mobilization among urban poor households in Kabul, Afghanistan. *Habitat International,* **32** (2), 248–260.

Kanyenze, G., Chitiyo, T., Mahere, T., Makwavarara, T., Mbire, P. & Moyo, E. (2003) *Giving Voice to the Unprotected Workers in the Informal Sector Economy in Africa: the Case of Zimbabwe.* Paper prepared for the ILO/SRO, Harare.

Kaseva, M.E. & Mbuligwe, S.E. (2005) Appraisal of solid waste collection following private sector involvement in Dar es Salaam city, Tanzania. *Habitat International,* **28** (2), 353–366.

Kassim, S.M. & Ali, M. (2006a) Solid waste collection by the private sector: households' perspective – findings from a study in Dar es Salaam city, Tanzania. *Habitat International,* **30** (4), 769–780.

Kassim, S.M. & Ali, M. (2006b) Solid waste collection following private sector involvement in Dar es Salaam city, Tanzania. *Habitat International,* **30** (4), 781–796.

Keivani, R., Parsa, A. & McGreal, S. (2001) Globalisation, institutional structures and real estate markets in central European cities. *Urban Studies,* **38** (13), 2457–2476.

Kiwasila, H. (2003) *Promoting Decent Work for the Poor in Solid Waste Management and Elimination of Child Labour.* Unpublished study undertaken for the ILO, Dar es Salaam.

Klink, J. (1999) The future is coming. Economic restructuring in the São Paulo fringe: the case of Diadema. *Habitat International,* **23** (3), 325–338.

Klink, J. (2003) O novo regionalismo a maneira do ABC: em busca de uma economia regional de aprendizagem. *Cadernos de Pesquisa CEBRAP,* São Paulo, Volumen 8.

Klink, J. (2006) *The Role of Local Authorities in Promoting Decent Work: Towards an Applied Research Agenda for the Construction and Urban Development Sector.* Working Paper No. 243, Sectoral Activities Programme, International Labour Organization, Geneva.

Klink, J. (2008) Cities, international labor migration and development: towards an alternative research agenda. *Habitat International,* **32** (2), 237–247.

Krishna, A. & Uphoff, N. (1999) *Mapping and Measuring Social Capital: a Conceptual and Empirical Study of Collective Action for Conserving and Developing Watersheds in Rajasthan, India.* The World Bank, Social Capital Initiative Working Paper No. 13.

Kucera, D. (ed.) (2007) *Qualitative Indicators of Labour Standards: Comparative Methods and Applications.* Social indicators research series, International Labour Organization, Geneva.

Kulaba, S. (1989) Local government and the management of urban services in Tanzania. In R.E. Stren & R.R. White (eds) *African Cities in Crisis: Managing Rapid Urban Growth.* Westview, Boulder, Colorado.

Kuruvilla, S. (2003) *Social Dialogue for Decent Work.* Discussion Paper No. 149. International Institute of Labour Studies, International Labour Organization, Geneva.

Kyessi, A.G. (2005) Community based urban water management in fringe neighbourhoods: the case of Dar es Salaam, Tanzania. *Habitat International,* **29** (1), 1–25.

Labour and Human Rights Centre (no date) *Labour and Employment Issues.* Labour and Human Rights Centre, Dar es Salaam.

Ladbury, S., Cotton, A. & Jennings, M. (2003) *Implementing Labour Standards in Construction: a Sourcebook.* Water, Engineering and Development Centre, WEDC, Loughborough University, Leicestershire.

Larangeira, A.A. (2003) *Estudo de Caso: Programa Santo André Mais Igual; Intervenções em Sacadura Cabral, Tamarutaca, Capuava e Quilombo II.* IBAM/CAIXA, Rio de Janeiro.

Lawrence, R.J., Paredes, M., Flückiger, Y., Lambert, C. & Werna, E. (2008) Promoting decent work in the construction sector: the role of local authorities. *Habitat International,* **32** (2), 160–171.

LEDRIZ (2005) *Do you Know Your Eight (8) Socio-Economic Rights?* Labour Economic Development Research Institute of Zimbabwe, Harare.

Levaggi, V. (2006) *Democracia y Trabajo Decente en América Latina.* Oficina Regional para América Latina y el Caribe, Organización Internacional del Trabajo, Lima.

Lima, Jr. J.M., López-Valcárcel, A. & Alves Dias L. (2004) *Segurança e Saúde no Trabalho na Indústria da Construção no Brasil.* Documento de Trabalho Numero 200, SAFEWORK, Organización Internacional del Trabajo, Brasília.

Lindsay, C. & McQuaid, R.W. (2004) Avoiding the 'McJobs': unemployed job seekers and attitudes to service work. *Work Employment and Society,* **18** (2), 297–319.

Loewenson, R. (2001a) Effects of globalization on working women. *African Newsletter on Occupational Health and Safety,* **11** (3), 66–67.

Loewenson, R. (2001b) Globalization and occupational health: a perspective from Southern Africa. *Bulletin of the World Health Organization,* **79**, 863–868.

Madhuku, L. (2001) *Gender Equality in Employment: the Legal Framework in the Case of Zimbabwe.* Discussion Paper No. 19. ILO Southern Africa Multidisciplinary Advisory Team (ILO/SAMAT), Harare.

Majale, M. (2008) Employment creation through participatory urban planning and slum upgrading: the case of Kitale, Kenya. *Habitat International,* **32** (2), 270–282.

Majid, N. (2001) *Economic Growth, Social Policy and Decent Work.* Macroeconomic and Development Policy Group, Employment Strategy Department, International Labour Organization, Geneva.

Mbiba, B. (1999) Urban property ownership and the maintenance of communal land rights in Zimbabwe. Unpublished Ph.D. thesis, University of Sheffield.

Mbiba, B. (2001) Communal land rights as state sanction and social control – a narrative. *Africa*, **71** (3), 426–448.

Mbiba, B. (2005) Zimbabwe's global citizens in 'Harare North': some preliminary observations. In R. Palmberg & M. Primorac (eds) *Skinning The Skunk – Facing Zimbabwean Futures*. Nordic Africa Institute, Uppsala, pp. 26–39.

McGreal, S., Parsa, A. & Keivani, R. (2002) Evolution of property investment markets in central Europe: opportunities and constraints. *Journal of Property Research*, **19** (3), 213–230.

Miller, S. & Cohen, M. (2008) *Cities without Jobs?* ILO Discussion Paper on Urban Employment, International Labour Organization, Geneva.

Ministério do Trabalho e Emprego (2001) *Relação Anual de Informações Sociais*. RAIS, Brasília.

Mitra, A. (2008) Social capital, livelihood and upward mobility. *Habitat International*, **32** (2), 261–269.

Mlinga, R.S. & Wells, J. (2002) Collaboration between formal and informal enterprises in the construction sector in Tanzania. *Habitat International*, **26**, 269–280.

Moore, J. (2007) *ABC Region: Local Government Responses to De-industrialization and its Related Impacts*. School of Community and Regional Planning, University of British Columbia, Canada.

Mucheche, H. (2005a) The dawn of a new era – analysis of the rights of women under the new labour act. *The Civil Servant*, **4**, 10–14.

Mucheche, H. (2005b) The published Labour Amendment Bill: what exactly is this bill all about and what are its intensions? *The Civil Servant*, **5**, **3** and **12**.

Musiolek, B. (2002) *Decent Work in the Informal Sector*. ILO In Focus Programme on Boosting Employment through Small Enterprise Development, International Labour Organization, Geneva.

Mutizwa-Mangiza, N.D. (1991) The organization and management of urban local authorities in Zimbabwe: a case study of Bulawayo. *Third World Planning Review*, **13** (4), 357–380.

Narasimha Reddy, D. (2005) Challenges of decent work in the globalizing world. *Indian Journal of Labour Economics*, **48** (1), 3–18.

Narayan, D. (1997) *Voices of the Poor: Poverty and Social Capital in Tanzania*. The World Bank, Environmentally and Socially Sustainable Development Studies and Monographs Series, No. 20.

Narayan, D. (1999) *Complementarity and Substitution: Social Capital, Poverty Reduction and the State*. The World Bank, Poverty Group.

National Bureau of Statistics (2004) *Employment and Earnings Survey, 2001: Analytical Report*. National Bureau of Statistics, Dar es Salaam.

N'Dow, W. (1997) An urbanizing world. In U. Kirdar (ed.) *Cities Fit for People*. United Nations, New York, pp. 27–49.

Ngware, S. (1999) *An Assessment of People's Participation in Decision Making with Reference to Local Authorities*. Paper delivered at a workshop to develop strategies for a national program of governance, 29–31 March, Arusha.

Nnkya, T.J. (1999) *Towards Strategic Planning in Tanzania: the Case of the Sustainable Dar es Salaam Project*. Paper presented to a regional workshop on local government strategic planning/management for Eastern and Southern Africa, May, Addis Ababa: http://www.tanzaniagateway.org

NSSA (1997) *The Pensioner*. Volume 2. National Social Security Authority, Harare.

NSSA (2000) *Annual Analysis of Preliminary Occupational Injury and Rehabilitation Statistics 1997 and 1998*. National Social Security Authority, Harare.

NSSA (2001) *Annual Report 2001*. National Social Security Authority, Harare.

NSSA (2005) *Social Security News*. Volume 2. National Social Security Authority, Harare.

Nyamu-Musembi, C. & Cornwall, A. (2004) *What is the 'Rights Based Approach' all About?* Perspectives from international development agencies, Institute of Development Studies Working Paper 234.

OECD (1998) *Local Management for More Effective Employment Policies*. Organization for Economic Cooperation and Development, Paris.

OECD (2003) *Combating Child Labour: a Review of Policies*. Organization for Economic Cooperation and Development, Paris.

OECD (2005) *OCDE Factbook: Economics, Environmental and Social Statistics*. Organization for Economic Cooperation and Development, Paris.

Paredes Gil, M., Lawrence, R.J., Flückiger, Y., Lambert, C. & Werna, E. (2008) Decent work in Santo André: results of a multi-method case study. *Habitat International*, **32** (2), 172–179.

Parsa, A., Keivani, R., Sim, L.L., Ong, S.E., Agarwal, A. & Younis, B. (2004) *Emerging Global Cities – Comparison of Singapore and the Cities of UAE*. RICS research publication, London.

Pasteur, D. (1992) *Good Local Government in Zimbabwe: a Case Study of Bulawayo and Mutare City Councils 1980–1991*. Development Administration Group, University of Birmingham.

Patel, H. & Chan, S. (2006) Zimbabwe's foreign policy: a conversation. *The Round Table: the Commonwealth Journal of International Affairs*, **95**, 175–190.

Pochmann, M. (2004), *Políticas de Inclusão Social, Resultados e Avaliação*. Cortez, São Paulo.

Potts, D. (2006) City life in Zimbabwe at a time of fear and loathing: urban planning, urban poverty and Operation Murambatsvina. In G. Myers & M. Murray (eds) *Cities in Contemporary Africa*. Palgrave, New York.

Prefeitura de Santo André (2002) *Santo André Mais Igual: Programa Integrado de Inclusão Social*. PSA, Santo André.

Prefeitura de Santo André (2004a) Educação & Trabalho. *Observatório da Educação*. Boletim Numero 2. DET, Santo André.

Prefeitura de Santo André (2004b) *Observatório Econômico*. Boletim Numero 3, Ano 2. PSA, Santo André.

Prefeitura de Santo André (2004c) Observatório da Educação do Trabalho. *Observatório Econômico*. Boletim Numero 4, Ano 2. PSA, Santo André.

Prefeitura de Santo André (2004d) *Observatório Econômico*. Boletim Numero 5, Ano 2. PSA, Santo André.

Prefeitura de Santo André & Commissaõ Européia (2004e) *Políticas de Inclusão Social de Santo André: A Experiência do Programa de Apoio às Populações Desfavorecidas*. PSA, Santo André.

Prefeitura de Santo André (2006a) *Observatório Econômico*. Boletim Numero 12, Ano 4. PSA, Santo André.

Prefeitura de Santo André (2006b) *Plano Municipal de Habitação*. PSA, Santo André.

Prefeitura de Santo André (2006c) *Programa Integrado de Qualificação (PIQ)*. Secretaria de Educaçaõ e Formação Professional, Departamento de Educação do Trabalhador, Santo André.

Radwan, S. (1997) The future of urban employment. In U. Kirdar (ed.) *Cities Fit for People*. United Nations, New York, pp. 318–325.

Raftopoulos, B. (2004) *Education and the Crisis in Zimbabwe*. The Cannon Collins Educational Trust for Southern Africa, London.

Rakodi, C. & Lloyd-Jones, T. (2002) *Urban Livelihoods: a People-centred Approach to Reducing Poverty*. Earthscan, London.

Ramalho, J.R. (1996) Labour, restructuring of production, and development. A point of view from Latin America. *The Ecumenical Review*, **48** (3), 369–378.

Ray, P.K. (2002) Decent work in the informal sector: training issues. *The Indian Journal of Labour Economics*, **45** (4), 1163–1173.

Reich, R.B. (2002) Challenge of decent work. *International Labour Review*, **141** (1–2), 115–122.

Rodgers, G. (2001) *Decent Work as a Development Objective*. Lecture to the Forty-second Annual Meeting of the Indian Society of Labour Economics, Madhya Pradesh, India.

Rodriguez, A., Tomeray, J. & Klink, J. (2001) Local empowerment through economic restructuring in Brazil: the case of the greater ABC region. *Geoforum*, **32**, 459–469.

Rukuni, M. & Eicher, C.K. (eds) (1994) *Zimbabwe's Agricultural Revolution*. University of Zimbabwe Publications, Harare.

Sachikonye, L.M. (2006) *The Impact of Operation Mutrambatsvina/Clean Up on the Working People of Zimbabwe*. A report prepared for the Labour and Economic Development Research Institute of Zimbabwe (LEDRIZ), Harare.

Sachs, I. (2004) Inclusive development and decent work for all. *International Labour Review*, **143** (1–2), 161–184.

Saith, A. (2004) *Social Protection, Decent Work and Development*. Discussion Paper No. 152. International Institute of Labour Studies, International Labour Organization, Geneva.

Salewi, K.W. (2006) *Re-engineering Public Private Partnerships (PPPs) in Municipal Infrastructure and Service Provision for Local Economic Development and Reduction of Urban Poverty: Lessons from Best Practice*. Paper presented at Engineers' Day conference, 24 March, Dar es Salaam.

Sassen, S. (1996) Service employment regimes and the new poverty. In E. Mingione (ed.) *Urban Poverty and the Underclass: a Reader*. Blackwell, Oxford.

Schiphorst, F.B. (2001) *Strength and Weakness: the Rise of the Zimbabwe Congress of Trade Unions and the Development of Labour Relations, 1980–1995*. University of Leiden, Leiden.

Sen, A.K. (1999) *Development as Freedom*. Nopf, New York.

Sen, A. (2000) Work and rights. *International Labour Review*, **139** (2), 129–139.

Sennet, R. (1998) *The Corrosion of Character: the Personal Consequence of Work in the New Capitalism*. Norton, London.

Seragelden, M., Kim, S. &. Wahba, S. (2000) *Decentralization and Urban Infrastructure Management Capacity*. Background paper for the Third Global Report on Human Settlements, UNCHS/Habitat, Nairobi.

Servais, J.M. (2004) Globalization and decent work policy: reflections upon a new legal approach. *International Labour Review*, **143** (1–2), 186–207.

Sheuya, S.A. (1996) *Practices and Opportunities for Employment Intensive and Labour Based Approaches in Urban Infrastructure Programmes: Case of Unplanned Settlements in Dar es Salaam, Tanzania.* Report for ILO/ASIST, Nairobi.

Sibanda, O. (2004) Indigenous economic empowerment in Zimbabwe: a focus on the construction Industry. Unpublished MBA dissertation, Open University, Bulawayo.

Spertini, S. & Denaldi, R. (2001) *As Possibilidades Efetivas de Regularização Fundiária em Núcleos de Favelas.* Anais do Seminário Internacional Gestão da Terra Urbana e Habitação de Interesse Social. PUCC, Campinas, São Paulo.

Srinivas, S. (2008) Urban labour markets in the twenty-first century: dualism, regulation and the role(s) of the State. *Habitat International*, **32** (2), 141–159

Standing, G. (2002) From people's security surveys to a decent work index. *International Labour Review*, **141** (4), 441–454.

Stren, R.E. (2000) *IDRC and the Management of Sustainable Urban Development in Latin America: Lessons Learnt and the Demands for Knowledge.* New approaches to urban governance in Latin America. Paper presented to the IDRC Conference, Montevideo.

Tajgman, D. & de Veen, J. (1998) *Employment-intensive Infrastructure Programmes: Labour Policies and Practices.* Development Policies Department, International Labour Organization, Geneva.

Takala, J. (2002) *Introductory Report: Decent Work – Safe Work.* Sixteenth World Congress on Health and Safety at Work, International Labour Organization, Geneva.

Taylor, J.E. (2006) *International Migration and Economic Development.* International Symposium on International Migration and Development, Population Division Department of Economic and Social Affairs, United Nations Secretariat, Turin.

The World Bank & Instituto de Pesquisa Econômica Aplicada (2002) *Brazil Jobs Reports*, No. 24408-BR, Vol. I & II, The World Bank Brazil Country Management Unit, Latin American and the Caribbean Region & Instituto de Pesquisa Econômica Aplicada, with the support of the Government of Brazil, Ministry of Labour and Employment, Brasilia.

Therkildsen, Ole (1998) *Challenges of Local Government Reform.* Paper presented at a national symposium on civil service reform in Tanzania, 15–16 January, Dar es Salaam.

Tipple, A.G. (2005) The place of home-based enterprises in the informal sector: evidence from Cochabamba, New Delhi, Surabaya and Pretoria. *Urban Studies*, **42** (4), 611–632.

Tournée, J. & van Esch, W. (2001) *Community Contracts in Urban Infrastructure Works: Practical Lessons from Experience.* Advisory Support, Information Services and Training for Employment-intensive Infrastructure Development (ASIST-Africa), International Labour Organization, Geneva.

Tripp, A.M. (1997) *Changing the Rules: the Politics of Liberalization and the Urban Informal Economy in Tanzania.* University of California Press, Berkeley and Los Angeles.

TUCTA (2004) *Trade Union Membership Profile in Tanzania.* Trade Union Congress of Tanzania, Dar es Salaam.

Turner, J. (1967) Barriers and channels for housing in modernizing countries. *Journal of the American Institute of Planners*, **33** (3), 10–21.

Turner, J. (1968) The squatter settlement: architecture that works. In M. Pidgeon & G. Bell (eds) *The Architecture of Democracy*. Special issue of *Architectural Design*, **8** (38), 38–52.

Turner, J. (1977) *Housing by People: Towards Autonomy in Building Environments*. Pantheon Books, New York.

Turner, J. & Fichter, R. (1972) *Freedom to Build*. Macmillan, New York.

UMP Regional Office for Latin America and the Caribbean, UN-Habitat Regional Office for Latin America and the Caribbean, Municipality of Santo André (2001) *Santo André City Development Strategy Report. City Development Strategies: Lessons From UMP/UN-Habitat Experiences*. Urban Management Programme Publications Series No. 29, UN-Habitat, Nairobi, pp. 129–152.

UNCHS (1996) *The Human Settlements Conditions of the World's Urban Poor*. United Nations Centre for Human Settlements, Nairobi, pp. 243–250.

UNCHS/Habitat – ILO (1989) *Improving Income and Housing: Employment Generation in Low-income Settlements*. United Nations Centre for Human Settlements, Nairobi.

UNCHS/Habitat – ILO (1995) *Shelter Provision and Employment Generation*. United Nations Centre for Human Settlements, International Labour Organization, Geneva, Nairobi.

UNCHS/Habitat (2006) *State of the World's Cities Report 2006/7*. United Nations Centre for Human Settlements, Nairobi.

United Nations Centre for Human Settlements & International Labour Office (1985) *Shelter Provision and Employment Generation*. UNCHS-ILO, Nairobi and Geneva.

UNDP (United Nations Development Programme) (1999) *Cities and Sustainable Human Development*. UNDP Policy Paper. United Nations Development Programme, New York.

UNDP (2000) *Human Development Report 2000: Human Rights and Human Development*. United Nations Development Programme, New York.

UNDP (2005) *Human Development Report 2005: International Cooperation at a Crossroads: Aid, Trade and Security in an Unequal World*. United Nations Development Programme, New York.

UNICEF (2005) *The State of the World's Children 2006: Excluded and Invisible*. United Nations Children's Fund, New York.

United Nations Volunteers (UNV) (2001) *Caring Cities: Volunteerism in Urban Development and the Role of the United Nations Volunteers Programme*. United Nations Volunteers, Bonn.

Uriarte, O.E. (2002) *Trabajo Decente y Formación Profesional*. Boletin 151 Cinterfor/International Labour Organization, Organización Internacional del Trabajo, Geneva.

URT (1993) *The Labour Force Survey 1990/91 (Tanzania mainland)*. Bureau of Statistics, Presidents Office, Planning Commission and Labour Department, Ministry of Labour and Youth Development, United Republic of Tanzania, Dar es Salaam.

URT (1997) *The National Employment Policy*. Ministry of Labour and Youth Development, United Republic of Tanzania, Dar es Salaam.

URT (1998) *Dar es Salaam: Informal Sector Survey, 1995.* The Planning Commission and the Ministry of Labour and Youth Development, United Republic of Tanzania, Dar es Salaam.

URT (2001) *Child Labour in Tanzania, Country Report 2000/01: IFF and Child Labour Survey.* ILO (IPEC), National Bureau of Statistics and Ministry of Labour, United Republic of Tanzania, Dar es Salaam.

URT (2002) *Integrated Labour Force Survey 2000/01 Analytical Report.* National Bureau of Statistics and Ministry of Labour, Youth Development and Sport, United Republic of Tanzania, Dar es Salaam.

URT (2003) *Household Budget Survey 2000/01.* National Bureau of Statistics, United Republic of Tanzania, Dar es Salaam.

URT (2005a) *The Economic Survey 2004.* President's Office, Planning and Privatization, United Republic of Tanzania, Dar es Salaam.

URT (2005b) *Local Government Capital Development Grant (LGCDG) System: Implementation and Operations Guide.* President's Office, Regional Administration and Local Government, United Republic of Tanzania, Dar es Salaam.

USAID (1996) *Gender Analysis of the Zimbabwe Housing Guaranty Program.* Prepared by Plan Inc, Zimbabwe Pvt (Ltd) for United States Agency for International Development Mission in Zimbabwe, Harare.

Van Empel, C. (2008) Social dialogue for urban employment: changing concepts and practices. *Habitat International*, **32** (2), 180–191.

Van Empel, C. & Werna, E. (2008) *Labour-oriented Participatory Approach as an Instrument of Urban Governance.* Paper written for and presented at the Twelfth General Conference of the European Association of Development Institutes, Geneva, 24–28 June.

Viloria-Williams, J. (2006) *Urban Community Upgrading – Lessons from the Past, Prospects for the Future.* World Bank Institute, Washington.

Watermeyer, R.B. (2006) Poverty reduction responses to the Millennium Development Goals. *The Structural Engineer*, **84** (9), **27**–34.

Wells, J. (1986) *The Construction Industry in Developing Countries: Alternative Strategies for Development.* Billing and Sons Limited, Worcester, UK.

Werna, E. (1997) *Labour Migration in the Construction Industry in Latin America and the Caribbean.* Sectoral Activities Programme, Industrial Activities Branch, International Labour Organization, Geneva.

Werna, E. (2000) *Combating Urban Inequalities: Challenges for Managing Cities in the Developing World.* Edward Elgar, Aldershot.

Werna, E. (2001) Shelter, employment and the informal city in the context of the present economic scene: implications for participatory governance. Jorge Hardoy Honorable Mention Paper, published in *Habitat International*, **25** (2), 209–227.

Werna, E. (2008) Labour in urban areas: an introduction. *Habitat International*, **32** (2), 137–140.

Werna, E., Harpham, T., Blue, I. & Goldstein, G. (1998) *Healthy City Projects in Developing Countries – an International Approach to Local Problems.* Earthscan, London.

Werna, E., Dzikus, A., Ochola, L. & Kumarasuryar, M. (1999) *Implementing the Habitat Agenda: Towards Child-friendly Cities.* Avebury, Aldershot.

Williams, D. (2007) *Participatory Approaches for Planning and Construction-related Assistance in Settlement Upgrading and Expansion: the Roles of Tripartite Actors and other Stakeholders*, Working Paper 255. Sectoral Activities Branch, International Labour Organization, Geneva.

World Bank (1983) *Labour-based Construction Programs, a Practical Guide for Planning and Management.* International Bank for Reconstruction and Development, Oxford University Press, Washington, DC.

World Bank (2005a) *World Development Report 2006: Equity and Development.* The World Bank and Oxford University Press, New York.

World Bank (2005b) *Exploring Partnerships Between Communities and Local Governments in Community Driven Development: a Framework.* World Bank Social Development Department, Washington, DC.

Yin, R. (1994) *Case Study Research: Design and Methods.* Sage Publications, Thousand Oaks, CA.

You, N. (2007) Sustainable for whom? The urban millennium and challenges for redefining the global development planning agenda. *Cities*, **11** (2), 214–220.

Zaaijer, M. (1998) *Urban Economic Restructuring and Local Institutional Response: the Case of Bulawayo.* IHS Project Paper No. UM 1, Rotterdam.

Zarka-Martres, M. & Guichard-Kelly, M. (2005) *Decent Work, Standards and Indicators.* Working Paper No. 58. Statistical Development and Analysis Group, Policy Integration Department, International Labour Organization, Geneva.

ZCATWU/ZCTU (no date) *Improved Conditions of Service in the New Labour Relations Act.* Briefing produced by Organizing Department of the Zimbabwe Congress of Trade Unions, Zimbabwe Construction and Allied Trades Workers' Union and Zimbabwe Congress of Trade Unions.

ZCTU (2000) *The 'Workers' driven' and 'Peoples-Centered' Development Process for Zimbabwe, Comprehensive Report for the Advocacy Programme Beyond ESAP Phase 1* (Compiled by T.F. Kondo). Journeyman Press, Zimbabwe Congress of Trade Unions, Harare.

ZCTU (2005a) *The Decent Work Agenda and The Role of Trade Unions – Zimbabwe Case.* Zimbabwe Congress of Trade Unions, Harare.

ZCTU (2005b) Position Paper on National Plan of Action on Employment Creation And Poverty Alleviation (unpublished electronic submission). Zimbabwe Congress of Trade Unions, Harare.

Zimbabwe Building Contractors Association. *Code of Conduct.* Zimbabwe Building Contractors Association (ZBCA), 46 Central Avenue, Harare, Zimbabwe (in force 2006).

ZUCWU Bulawayo Branch (2006) *Annual General Meeting Chairman's Report.* Zimbabwe Urban Councils Workers' Union, Bulawayo.

Internet references

Beijing Declaration and Platform for Action, Fourth World Conference on Women (15 September 1995) (consulted on 2 May 2006)
http://www.un.org/womenwatch/daw/beijing/platform/declar.htm

Convention on the Elimination of All Forms of Discrimination against Women (consulted on 2 May 2006)
http://www.un.org/womenwatch/daw/cedaw/text/econvention.htm#article11

Convention on the Rights of the Child (consulted on 4 May 2006)
http://www.unhchr.ch/html/menu3/b/k2crc.htm

Declaration on the World Summit for Social Development (consulted on 3 April 2006)
http://www.un.org/documents/ga/conf166/aconf166-9sp.htm

Departamento Intersindical de Estatística e Estudos Sócio-econômicos (consulted on 15 September 2006)
http://www.dieese.org.br

Freedom House Civil Liberty Index (consulted on 13 April 2006)
http://www.freedomhouse.org

Fundação Sistema Estadual de Análise de Dados (consulted on 15 September 2006)
http://www.seade.gov.br

ILO (International Labour Organization) Constitution (last consultation May 2008)
http://www.ilo.org/public/english/about/iloconst.htm#a19

ILO Convention No. 29 Forced Labour Convention, 1930 (consulted on 22 March 2006)
http://www.ilo.org/ilolex/cgi-lex/convde.pl?C029

ILO Convention No. 100 Equal Remuneration Convention, 1951 (consulted on 22 March 2006)
http://www.ilo.org/ilolex/cgi-lex/convde.pl?C100

ILO Convention No. 111 Discrimination (Employment and Occupation) Convention, 1958 (consulted on 22 March 2006)
http://www.ilo.org/ilolex/cgi-lex/convde.pl?C111

ILO Convention No. 122 Employment Policy Convention 1964 (consulted on 22 March 2006)
http://www.ilo.org/ilolex/cgi-lex/convde.pl?C122

ILO Convention No. 138 Minimum Age Convention, 1973 (consulted on 22 March 2006)
http://www.ilo.org/ilolex/cgi-lex/convde.pl?C138

ILO Convention No. 167 Safety and Health in Construction Convention, 1988 (consulted on 22 March 2006)
http://www.ilo.org/ilolex/english/convdisp1.htm

ILO Convention No. 182 Worst Forms of Child Labour Convention, 1999 (consulted on 22 March 2006)
http://www.ilo.org/ilolex/cgi-lex/convde.pl?C182

ILO International Labour Conference, Eighty-sixth Session, June 1998. Declaration on fundamental principles and rights at work (consulted on 22 March 2006)
http://www.ilo.org/public/english/standards/relm/ilc/ilc86/com-dtxt.htm

ILO Declaration on Fundamental Principles and Rights at Work (consulted on 3 April 2006) http://www.ilo.org/dyn/declaris/DECLARATIONWEB.static_jump?var_language=EN&var_pagename=DECLARATIONTEXT

ILO LABORSTA Internet (consulted on 18 May 2006)
http://laborsta.ilo.org

ILO website, Social Dialogue, Sectoral activities, Construction (last consultation May 2008)
http://www.ilo.org/public/english/dialogue/sector/sectors/constr.htm

ILO website, International Programme on the Elimination of Child Labour (IPEC) (last consultation June 2008)
http://www.ilo.org/ipecinfo

International Covenant on Economic, Social and Cultural Rights (consulted on 3 April 2006)
http://www.unhchr.ch/html/menu3/b/a_cescr.htm

International Relations Center, Rights web (consulted on 15 August 2006)
http://rightweb.irc-online.org

Instituto Brasileiro de Geografia e Estatística: Data at the Municipal Level: Santo André (consulted on 7 August 2006)
http://www.ibge.gov.br

Instituto Ecuatoriano de Seguridad Social (IESS) (consulted on 21 August 2006)
http://www.iess.gov.ec

Ministerio Brasiliero do Trabalho e Emprego (consulted on 7 August 2006)
http://www.mte.gov.br

Prefeitura de Santo André, Brasil (consulted on 26 July 2006)
http://www.santoandre.sp.gov.br

Relação Anual de Informações Sociais, Brasil (consulted on 15 September 2006)
http://www.rais.gov.br

The Economist Intelligence Unit: Brazil country profile 2005 (consulted on 26 July 2006)
http://www.eiu.com

The Herald, Harare (last consulted 3 May 2006)
http://www.herald.co.zw

UN Habitat Conference on Human Settlements 1996. 'City summit end with leaders' commitment to improve living standards' (consulted on 18 May 2006)
http://www.un.org/Conferences/habitat/eng-pres/3/habist25.htm

Universal Declaration of Human Rights (consulted on 3 April 2006)
http://www.un.org/Overview/rights.html

United Nations Department of Economic and Social Affairs, Division for Sustainable Development, Agenda 21, Chapter 7: Promoting sustainable human settlement development (consulted on 8 May 2006)
http://www.un.org/esa/sustdev/documents/agenda21/english/agenda21chapter7.htm

United Nations, United Nations Earth Summit+5 Success Stories (consulted on 21 March 2007)
http://www.un.org/esa/earthsummit/recycle.htm.

University of British Columbia, New Public Consortia for Metropolitan Governance, Brazil (consulted on 30 May 2008)
http://www.chs.ubc.ca/our-projects.html

Index